Catalytic Hydrogen Generation and Use for Production of Fuels

Catalytic Hydrogen Generation and Use for Production of Fuels

Editor

Dmitri A. Bulushev

MDPI • Basel • Beijing • Wuhan • Barcelona • Belgrade • Manchester • Tokyo • Cluj • Tianjin

Editor
Dmitri A. Bulushev
Boreskov Institute of Catalysis
Russia

Editorial Office
MDPI
St. Alban-Anlage 66
4052 Basel, Switzerland

This is a reprint of articles from the Special Issue published online in the open access journal *Energies* (ISSN 1996-1073) (available at: https://www.mdpi.com/journal/energies/special_issues/Catalytic_Hydrogen_Generation_Use).

For citation purposes, cite each article independently as indicated on the article page online and as indicated below:

LastName, A.A.; LastName, B.B.; LastName, C.C. Article Title. *Journal Name* **Year**, *Volume Number*, Page Range.

ISBN 978-3-0365-4673-5 (Hbk)
ISBN 978-3-0365-4674-2 (PDF)

© 2022 by the authors. Articles in this book are Open Access and distributed under the Creative Commons Attribution (CC BY) license, which allows users to download, copy and build upon published articles, as long as the author and publisher are properly credited, which ensures maximum dissemination and a wider impact of our publications.

The book as a whole is distributed by MDPI under the terms and conditions of the Creative Commons license CC BY-NC-ND.

Contents

About the Editor . . . . **vii**

Preface to "Catalytic Hydrogen Generation and Use for Production of Fuels" . . . . **ix**

Dmitri A. Bulushev
Advanced Catalysis in Hydrogen Production from Formic Acid and Methanol
Reprinted from: *Energies* **2021**, *14*, 6810, doi:10.3390/en14206810 . . . . **1**

Dmitri A. Bulushev
Progress in Catalytic Hydrogen Production from Formic Acid over Supported Metal Complexes
Reprinted from: *Energies* **2021**, *14*, 1334, doi:10.3390/en14051334 . . . . **7**

Miriam Navlani-García, David Salinas-Torres and Diego Cazorla-Amorós
Hydrolytic Dehydrogenation of Ammonia Borane Attained by Ru-Based Catalysts: An Auspicious Option to Produce Hydrogen from a Solid Hydrogen Carrier Molecule
Reprinted from: *Energies* **2021**, *14*, 2199, doi:10.3390/en14082199 . . . . **21**

Arina N. Suboch and Olga Y. Podyacheva
Pd Catalysts Supported on Bamboo-Like Nitrogen-Doped Carbon Nanotubes for Hydrogen Production
Reprinted from: *Energies* **2021**, *14*, 1501, doi:10.3390/en14051501 . . . . **41**

Vladimir V. Chesnokov, Pavel P. Dik and Aleksandra S. Chichkan
Formic Acid as a Hydrogen Donor for Catalytic Transformations of Tar
Reprinted from: *Energies* **2020**, *13*, 4515, doi:10.3390/en13174515 . . . . **55**

Ekaterina Matus, Olga Sukhova, Ilyas Ismagilov, Mikhail Kerzhentsev, Olga Stonkus and Zinfer Ismagilov
Hydrogen Production through Autothermal Reforming of Ethanol: Enhancement of Ni Catalyst Performance via Promotion
Reprinted from: *Energies* **2021**, *14*, 5176, doi:10.3390/en14165176 . . . . **67**

About the Editor

Dmitri A. Bulushev

Dmitri A. Bulushev graduated from the Novosibirsk State University (Russia) in 1983 and received his PhD degree in chemistry (chemical kinetics and catalysis) from the Boreskov Institute of Catalysis (Russia) in 1991. From 1995 to 2007, he was a researcher at the University of Ghent (Belgium) and EPFL (Switzerland) before serving as a senior research fellow at the University of Limerick (Ireland). Since 2014, he has been working as a senior researcher at the Boreskov Institute of Catalysis. He has published more than 75 peer-reviewed papers, which have received over 3000 citations. He supervised two PhD students. Dmitri is a reviewer for more than 50 international journals and 10 science-funding organizations. Since 2019, he has been a member of the Editorial Board of the MDPI journal Energies. His research interests include hydrogen production, conversion of biomass to valuable chemicals, studying factors determining the activity and selectivity of catalysts, and the application of novel carbon materials as catalyst supports.

Preface to "Catalytic Hydrogen Generation and Use for Production of Fuels"

Hydrogen is considered as a fuel for the future. Catalytic approaches to produce hydrogen involve dehydrogenation, gasification, water–gas shift, and steam and dry reforming reactions. Recent studies have considered the utilization of new sources of hydrogen, such as biomass, as well as liquid organic and solid hydrogen carriers. Photo- and electrocatalytic methods for hydrogen production have become important. Hydrogen is also intensively used for the synthesis of fuels via catalysis. Active, selective, and stable supported catalysts are needed for all these processes.

The aim of the book "Catalytic Hydrogen Generation and Use for Production of Fuels" is to discuss the field of catalytic hydrogen production and its application in the synthesis of fuels. Included are 2 review papers and 3 experimental papers related to hydrogen generation and hydrogen use in fuel production. Dr. D. Bulushev and Dr. M. Navlani-García et al. present reviews related to hydrogen production from formic acid over supported metal complexes and from ammonia borane over Ru-based catalysts, respectively. Dr. A. Suboch and Dr. O. Podyacheva studied hydrogen production over Pd catalysts supported on N-doped carbon nanotubes. Dr. V. Chesnokov et al. successfully used formic acid as a hydrogen source and a Ni–Mo-based catalyst to upgrade tar. Dr. E. Matus et al. applied different bimetallic Ni-containing catalysts to produce hydrogen by autothermal reforming of ethanol.

This book is potentially useful for specialists in catalysis and nanomaterials as well as for graduate students studying chemistry and chemical engineering. The reported results can be applied in the development of catalysts for hydrogen production from different liquid organic hydrogen carriers. We acknowledge the contribution of MDPI in publishing this book.

Dmitri A. Bulushev
Editor

Editorial

Advanced Catalysis in Hydrogen Production from Formic Acid and Methanol

Dmitri A. Bulushev

Department of Nontraditional Catalytic Processes, Boreskov Institute of Catalysis SB RAS, 630090 Novosibirsk, Russia; dmitri.bulushev@catalysis.ru

Abstract: The Special Issue of the *Energies* journal related to the hydrogen production from formic acid decomposition was published recently by MDPI. This Editorial note contains a short analysis of the papers published in this Special Issue and some historical information connected to this reaction.

Keywords: formic acid; hydrogen production; catalysts

Hydrogen production from different hydrogen carriers is an important topic that has been examined in many studies, as hydrogen is considered to be a clean fuel for the future, giving only water as a product. At the same time, efforts for its storage and transportation may encounter serious difficulties related to safety. Hence, it is necessary to develop liquid and solid hydrogen carriers that will allow the safe storage/transportation of hydrogen and its liberation at mild conditions using catalysts. Therefore, many recent studies in this field have focused on the development of efficient catalysts for hydrogen production from different compounds.

MDPI published the Special Issue "Advanced Catalysis in Hydrogen Production from Formic Acid and Methanol" in the *Energies* journal and as a separate book, for which I served as Guest Editor. Formic acid and methanol are liquid organic hydrogen carriers (LOHCs) that can be produced from biomass [1] or by CO_2 hydrogenation [2,3]. The Special Issue included five invited research papers and two invited reviews. Unfortunately, no papers related to the production of hydrogen from methanol were presented; hence, the issue was focused on different aspects of hydrogen production from formic acid using heterogeneous and homogeneous catalysts (Figure 1).

Studies of the catalytic decomposition of formic acid have more than 150 years of history and have contributed significantly to the science of catalysis. Formic acid decomposes through two routes, giving hydrogen and carbon dioxide (1, dehydrogenation) and carbon monoxide and water (2, dehydration).

$$HCOOH \rightarrow CO_2 + H_2 \tag{1}$$

$$HCOOH \rightarrow CO + H_2O \tag{2}$$

The German chemist Döbereiner was the first to report the dehydration of formic acid by heating it with anhydrous sulfuric acid in 1821 [4]. This reaction was studied in detail at the beginning of the last century. Some heterogeneous catalysts were tested. If the goal is to produce pure hydrogen, dehydration should be eliminated using catalysts. However, dehydration is important for applications where formic acid is applied as a CO source—for example, synthesis gas production [1,5].

The French chemist Marcellin Berthelot was the first to report the dehydrogenation of formic acid in 1864 [6]. He conducted the decomposition of formic acid vapor over 15 g of Pt black powder at 443 and 533 K and showed that equal amounts of CO_2 and H_2 were formed. Later, the Nobel Prize Winner in Chemistry (1912) Paul Sabatier and his co-author A. Mailhe showed that some heterogeneous catalysts such as Pt, Pd, Ni, Cu, Cd, SnO_2, and

ZnO were active in dehydrogenation reaction, while oxides of Ti, Al, Si, Zr, U, and W were active in dehydration reaction [7]. Formaldehyde was also formed sometimes. It should be mentioned that Sabatier was an assistant in the laboratory of Berthelot and received his Doctor of Science degree under his supervision in 1880 [8].

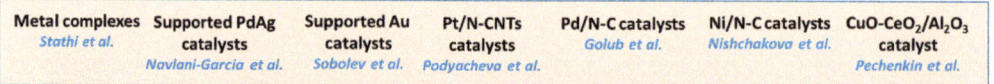

Figure 1. Catalysts for hydrogen production from formic acid decomposition considered in the Special Issue.

In 20th century, formic acid decomposition was studied as a model reaction used for the establishment of the bases of catalysis and for the development of spectroscopic methods. Later, Density Functional Theory (DFT) studies were performed to elucidate the mechanism of the reaction.

Formic acid contains a considerable amount of hydrogen (53 g L^{-1}). Since 2008–2010, it has been considered as a substance that could be used for hydrogen storage. Homogeneous catalysts have been used intensively for liquid-phase formic acid decomposition and for formic acid production from CO_2 and H_2 [9–14]. At the same time, heterogeneous catalysts have also been used for liquid-phase [15] and gas-phase [16,17] reactions with the purpose of producing CO-free hydrogen efficiently. These catalysts possess a significant advantage with respect to homogeneous catalysts, allowing to separate them easily from the reaction mixture [18,19].

There are several ways to achieve efficient hydrogen production from formic acid using catalysts. These involve support modification, metal size optimization, the alloying of active metal with another metal, and promotion with basic compounds containing alkali metals or amines. All these methods are considered in this Special Issue. The reviews featured to discuss liquid-phase formic acid decomposition over bimetallic (PdAg) [20], molecular (Ru, Rh, Ir, Fe, Co, and others) [21], and heterogenized molecular Fe catalysts [21]. The gas-phase reaction is studied over highly dispersed Pd [22], Pt [23], Au [24], Cu [25], and Ni [26]-supported catalysts.

Stathi et al. [21], in their review, discuss the features of the production of hydrogen from formic acid over a set of homogeneous catalysts represented by metal complexes. The mechanism of the liquid phase reaction was analyzed based on DFT studies. The authors indicated the importance of the deprotonation of formic acid as a first step of the reaction. They also outlined other important factors: the nature of the solvent and ligand, temperature and pH. Moreover, they discussed the continuous operation of hydrogen production from formic acid and showed some examples where immobilized molecular catalysts demonstrated comparable properties to nonimmobilized catalysts. This latter feature is important for the creation of immobilized molecular catalysts [18].

Navlani-García et al. [20], in their review, discussed the properties of PdAg catalysts in the production of hydrogen from formic acid. The interest in these catalysts is based on their high performance in liquid-phase reactions [15,27,28]. The authors ascribed the enhancement displayed by PdAg catalysts as compared to their monometallic counterparts to several effects, such as the formation of electron-rich Pd species and increased resistance to CO poisoning. Additionally, the authors considered the photocatalytic decomposition of formic acid. They also concluded that very little is known about the performance of the catalysts in highly concentrated formic acid solutions. It would be interesting to study the decomposition of gas-phase formic acid over highly dispersed PdAg catalysts.

The nature of the catalyst's support plays an important role in the reaction when the metal is highly dispersed. Sobolev et al. [24] studied the gas-phase formic acid decomposition over Au catalysts with a mean size of 1.6–2.4 nm supported on SiO_2, TiO_2, and Al_2O_3 and compared their catalytic properties with the properties of the corresponding supports. They concluded that the undesirable dehydration pathway was provided by the acid–base properties of the support. Thus, the selectivity in hydrogen production was found to be very low for the Au/TiO$_2$ catalyst (<20%), but it was close to 100% for the Au/Al$_2$O$_3$ catalyst. At the same time, the conversion of formic acid over the former catalyst was entirely provided by the support, since the conversion temperature dependences were almost the same. Earlier, we also indicated the importance of the nature of the support for highly dispersed Au catalysts and this reaction [29,30]. Gold supported on an N-doped porous carbon support showed a higher hydrogen yield than gold supported on silica or alumina.

In this Special Issue, the effect of carbon doping with nitrogen species was analyzed in detail for Pt [23], Pd [22], and Ni [26] catalysts. Nitrogen was inserted into the structure of carbon supports by direct synthesis using N-containing precursors [23,26] or by the post-deposition of N-containing precursors on the surface of the carbon support with optimized properties [22]. The latter method allows the scaling of the synthesis of N-doped carbon and could be useful for industrial applications.

It is important that the N-doping of the carbon support exerts a significant promotional effect. One of the reasons for the high activity of N-doped catalysts is the improvement in metal dispersion and formation of single-atom metal sites stabilized by pyridinic N species present in the support [19]. For example, Podyacheva et al. [23] demonstrated the higher activity of single Pt atoms supported on N-doped carbon nanotubes or carbon nanofibers as compared to Pt atoms on the surface of Pt nanoparticles. A correlation with the concentration of pyridinic N atoms was shown. In a recent paper [31], this group demonstrated the same effects for Pd catalysts.

Golub et al. [22] used different N-containing precursors (bypiridine, phenanthroline, and melamine) to deposit over a graphitic carbon support (Sibunit). The best performance was discovered for Pd catalysts, for which melamine was used as the N-precursor. The deposition of melamine followed by pyrolysis led to an increase in the activity and selectivity and to a decrease in the apparent activation energy. Thus, the catalytic activity of the N-doped Pd catalyst was higher by a factor of 4 than that of the Pd catalyst supported on the N-free carbon support at 373 K.

The properties of non-noble metal catalysts (Ni and Cu) were also discussed in this Special Issue. Normally, the activity of these catalysts is lower than that of noble metals such as Pt and Pd, but their price is also lower and this is important. Nishchakova et al. [26] studied Ni catalysts supported on N-doped and N-free porous carbon prepared at different temperatures. A temperature equal to 1073 K was found to be optimal for the catalytic activity. The N-doped Ni catalysts possessed a high stability in the formic acid decomposition reaction and a slightly higher activity than the N-free catalyst, with a mean particle size of 3.9 nm. A further study in this field [32] performed by the same group demonstrated that the used N-doped catalyst was a single-atom catalyst with active Ni-N$_4$ sites.

Pechenkin et al. [25] used a Cu-containing catalyst supported on a CeO$_2$/γ-Al$_2$O$_3$ support. The gas-phase decomposition of formic acid was studied in detail. The authors showed a very high yield of hydrogen equal to 98% at temperatures of 473–573 K. This yield was higher than those obtained for methanol and dimethoxymethane steam reforming reactions. The catalyst was stable in the reaction conditions used.

Therefore, key problems related to catalytic activity in hydrogen production from formic acid were discussed in this Special Issue. Interestingly, the amounts of noble metals in the Pt [23] and Pd [22] catalysts discussed in this issue were by a factor of 10^5 lower than those used in the experiments carried out by Berthelot [6] at similar temperatures. This reflects the progress made in the development of catalysts during the last 150 years, indicating that modern catalysts are significantly more active.

Finally, the results reported in this Special Issue can be applied for the development of catalysts for hydrogen production not only from formic acid, but also from other organic hydrogen carriers.

Funding: The preparation of this note is supported by the Russian Science Foundation (grant 17-73-30032).

Conflicts of Interest: The authors declare no conflict of interest.

References

1. Bulushev, D.A.; Ross, J.R.H. Towards sustainable production of formic acid. *ChemSusChem* **2018**, *11*, 821–836. [CrossRef]
2. Álvarez, A.; Bansode, A.; Urakawa, A.; Bavykina, A.V.; Wezendonk, T.A.; Makkee, M.; Gascon, J.; Kapteijn, F. Challenges in the greener production of formates/formic acid, methanol, and DME by heterogeneously catalyzed CO_2 hydrogenation processes. *Chem. Rev.* **2017**, *117*, 9804–9838. [CrossRef]
3. Bulushev, D.A.; Ross, J.R. Heterogeneous catalysts for hydrogenation of CO_2 and bicarbonates to formic acid and formates. *Catal. Rev.* **2018**, *60*, 566–593. [CrossRef]
4. Schierz, E.R. The decomposition of formic acid by sulfuric acid1. *J. Am. Chem. Soc.* **1923**, *45*, 447–455. [CrossRef]
5. Albert, J.; Jess, A.; Kern, C.; Pöhlmann, F.; Glowienka, K.; Wasserscheid, P. Formic acid-based Fischer–Tropsch synthesis for green fuel production from wet waste biomass and renewable excess energy. *ACS Sustain. Chem. Eng.* **2016**, *4*, 5078–5086. [CrossRef]
6. Berthelot, M. Notes sur la decomposition de l'acide formique. *C. R. Chim.* **1864**, *59*, 901.
7. Sabatier, P.; Mailhe, A. Sur la decomposition catalytique de l'acide formique. *C. R. Chim.* **1911**, *152*, 1212–1215.
8. Fechete, I. Paul Sabatier—The father of the chemical theory of catalysis. *C. R. Chim.* **2016**, *19*, 1374–1381. [CrossRef]
9. Joó, F. Breakthroughs in hydrogen storage-formic acid as a sustainable storage material for hydrogen. *ChemSusChem* **2008**, *1*, 805–808. [CrossRef]
10. Boddien, A.; Loges, B.; Junge, H.; Beller, M. Hydrogen generation at ambient conditions: Application in fuel cells. *ChemSusChem* **2008**, *1*, 751–758. [CrossRef]
11. Fukuzumi, S. Bioinspired energy conversion systems for hydrogen production and storage. *Eur. J. Inorg. Chem.* **2008**, *2008*, 1351–1362. [CrossRef]
12. Enthaler, S. Carbon dioxide—The hydrogen-storage material of the future? *ChemSusChem* **2008**, *1*, 801–804. [CrossRef] [PubMed]
13. Himeda, Y. Highly efficient hydrogen evolution by decomposition of formic acid using an iridium catalyst with 4,4′-dihydroxy-2,2′-bipyridine. *Green Chem.* **2009**, *11*, 2018–2022. [CrossRef]
14. Fellay, C.; Dyson, P.; Laurenczy, G. A viable hydrogen-storage system based on selective formic acid decomposition with a ruthenium catalyst. *Angew. Chem. Int. Ed.* **2008**, *47*, 3966–3968. [CrossRef] [PubMed]
15. Zhou, X.; Huang, Y.; Xing, W.; Liu, C.; Liao, J.; Lu, T. High-quality hydrogen from the catalyzed decomposition of formic acid by Pd–Au/C and Pd–Ag/C. *Chem. Commun.* **2008**, *30*, 3540–3542. [CrossRef] [PubMed]
16. Bulushev, D.; Beloshapkin, S.; Ross, J.R. Hydrogen from formic acid decomposition over Pd and Au catalysts. *Catal. Today* **2010**, *154*, 7–12. [CrossRef]
17. Solymosi, F.; Koós, Á.; Liliom, N.; Ugrai, I. Production of CO-free H_2 from formic acid. A comparative study of the catalytic behavior of Pt metals on a carbon support. *J. Catal.* **2011**, *279*, 213–219. [CrossRef]
18. Bulushev, D. Progress in catalytic hydrogen production from formic acid over supported metal complexes. *Energies* **2021**, *14*, 1334. [CrossRef]
19. Bulushev, D.A.; Bulusheva, L.G. Catalysts with single metal atoms for the hydrogen production from formic acid. *Catal. Rev.* **2021**, 1–40. [CrossRef]
20. Navlani-García, M.; Salinas-Torres, D.; Cazorla-Amorós, D. Hydrogen production from formic acid attained by bimetallic heterogeneous PdAg catalytic systems. *Energies* **2019**, *12*, 4027. [CrossRef]
21. Stathi, P.; Solakidou, M.; Louloudi, M.; Deligiannakis, Y. From homogeneous to heterogenized molecular catalysts for H_2 production by formic acid dehydrogenation: Mechanistic aspects, role of additives, and co-catalysts. *Energies* **2020**, *13*, 733. [CrossRef]
22. Golub, F.S.; Beloshapkin, S.; Gusel'Nikov, A.V.; Bolotov, V.A.; Parmon, V.N.; Bulushev, D.A. Boosting hydrogen production from formic acid over Pd catalysts by deposition of N-containing precursors on the carbon support. *Energies* **2019**, *12*, 3885. [CrossRef]
23. Podyacheva, O.; Lisitsyn, A.; Kibis, L.; Boronin, A.; Stonkus, O.; Zaikovskii, V.; Suboch, A.; Sobolev, V.; Parmon, V. Nitrogen doped carbon nanotubes and nanofibers for green hydrogen production: Similarities in the nature of nitrogen species, metal–nitrogen interaction, and catalytic properties. *Energies* **2019**, *12*, 3976. [CrossRef]
24. Sobolev, V.; Asanov, I.; Koltunov, K. The role of support in formic acid decomposition on gold catalysts. *Energies* **2019**, *12*, 4198. [CrossRef]
25. Pechenkin, A.; Badmaev, S.; Belyaev, V.; Sobyanin, V. Production of hydrogen-rich gas by formic acid decomposition over $CuO\text{-}CeO_2/\gamma\text{-}Al_2O_3$ Catalyst. *Energies* **2019**, *12*, 3577. [CrossRef]
26. Nishchakova, A.D.; Bulushev, D.A.; Stonkus, O.A.; Asanov, I.P.; Ishchenko, A.V.; Okotrub, A.V.; Bulusheva, L.G. Effects of the carbon support doping with nitrogen for the hydrogen production from formic acid over Ni catalysts. *Energies* **2019**, *12*, 4111. [CrossRef]
27. Tedsree, K.; Li, T.; Jones, S.C.; Chan, C.W.A.; Yu, K.M.K.; Bagot, P.; Marquis, E.; Smith, G.D.W.; Tsang, S.C.E. Hydrogen production from formic acid decomposition at room temperature using a Ag–Pd core–shell nanocatalyst. *Nat. Nanotechnol.* **2011**, *6*, 302–307. [CrossRef]

28. Navlani-García, M.; Mori, K.; Nozaki, A.; Kuwahara, Y.; Yamashita, H. Screening of carbon-supported PdAg nanoparticles in the hydrogen production from formic acid. *Ind. Eng. Chem. Res.* **2016**, *55*, 7612–7620. [CrossRef]
29. Zacharska, M.; Chuvilin, A.L.; Kriventsov, V.V.; Beloshapkin, S.; Estrada, M.; Simakov, A.; Bulushev, D.A. Support effect for nanosized Au catalysts in hydrogen production from formic acid decomposition. *Catal. Sci. Technol.* **2016**, *6*, 6853–6860. [CrossRef]
30. Bulushev, D.A.; Sobolev, V.I.; Pirutko, L.V.; Starostina, A.V.; Asanov, I.P.; Modin, E.; Chuvilin, A.L.; Gupta, N.; Okotrub, A.V.; Bulusheva, L.G. Hydrogen production from formic acid over Au catalysts supported on carbon: Comparison with Au catalysts supported on SiO_2 and Al_2O_3. *Catalysts* **2019**, *9*, 376. [CrossRef]
31. Suboch, A.; Podyacheva, O. Pd catalysts supported on bamboo-like nitrogen-doped carbon nanotubes for hydrogen production. *Energies* **2021**, *14*, 1501. [CrossRef]
32. Bulushev, D.A.; Nishchakova, A.D.; Trubina, S.V.; Stonkus, O.A.; Asanov, I.P.; Okotrub, A.V.; Bulusheva, L.G. $Ni-N_4$ sites in a single-atom Ni catalyst on N-doped carbon for hydrogen production from formic acid. *J. Catal.* **2021**, *402*, 264–274. [CrossRef]

Review

Progress in Catalytic Hydrogen Production from Formic Acid over Supported Metal Complexes

Dmitri A. Bulushev

Laboratory of Catalytic Methods of Solar Energy Transformation, Boreskov Institute of Catalysis, SB RAS, 630090 Novosibirsk, Russia; dmitri.bulushev@catalysis.ru

Abstract: Formic acid is a liquid organic hydrogen carrier giving hydrogen on demand using catalysts. Metal complexes are known to be used as efficient catalysts for the hydrogen production from formic acid decomposition. Their performance could be better than those of supported catalysts with metal nanoparticles. However, difficulties to separate metal complexes from the reaction mixture limit their industrial applications. This problem can be resolved by supporting metal complexes on the surface of different supports, which may additionally provide some surface sites for the formic acid activation. The review analyzes the literature on the application of supported metal complexes in the hydrogen production from formic acid. It shows that the catalytic activity of some stable Ru and Ir supported metal complexes may exceed the activity of homogeneous metal complexes used for deposition. Non-noble metal-based complexes containing Fe demonstrated sufficiently high performance in the reaction; however, they can be poisoned by water present in formic acid. The proposed review could be useful for development of novel catalysts for the hydrogen production.

Keywords: formic acid decomposition; hydrogen; biomass; metal complex; heterogeneous catalyst; ruthenium; iridium; iron

Citation: Bulushev, D.A. Progress in Catalytic Hydrogen Production from Formic Acid over Supported Metal Complexes. *Energies* **2021**, *14*, 1334. https://doi.org/10.3390/en14051334

Academic Editors: Ricardo Bessa and Wei-Hsin Chen

Received: 23 January 2021
Accepted: 22 February 2021
Published: 1 March 2021

Publisher's Note: MDPI stays neutral with regard to jurisdictional claims in published maps and institutional affiliations.

Copyright: © 2021 by the author. Licensee MDPI, Basel, Switzerland. This article is an open access article distributed under the terms and conditions of the Creative Commons Attribution (CC BY) license (https://creativecommons.org/licenses/by/4.0/).

1. Introduction

Hydrogen is mainly used for ammonia synthesis and the petrochemical industry. Its traditional production involves non-renewable sources and processes giving a significant emission of carbon dioxide leading to global warming. Among these processes are steam reforming of natural gas and gasification of coal performed at very high temperatures (>900 K). Recently, the International Energy Agency reported that the hydrogen production reached 75 mln of tons and that it was accompanied by emission of 830 mln tons of CO_2 [1]. Global demand for hydrogen increases from year to year accompanying by an increase of the carbon dioxide emissions.

Despite hydrogen is a clean energy carrier its safe transportation and storage are rather complicated. Liquid organic hydrogen carriers (LOHCs) are used for safe storage and transportation of hydrogen [2,3]. They can be produced from biomass or CO_2 thus avoiding the effect of the evolved CO_2 for global warming. Formic acid (HCOOH) is an example of such a LOHC. It contains 53.4 g L^{-1} hydrogen (4.4 wt %), which is by a factor of 2 higher than the content of compressed hydrogen at 350 bar at the same volume. This amount corresponds to the energy density of 2.1 kWh L^{-1}. In contrast to hydrogen, formic acid can be easily transported and stored and its application is much safer. An important feature of using formic acid is that it can be produced by catalytic hydrolysis/oxidation of biomass with high yields at low temperatures (<423 K) [4–6]. Hydrogen can be released from formic acid using catalysts at even lower temperatures (Figure 1). Thus, transformation of biomass to hydrogen through formic acid could be considered as an efficient route, because direct gasification of biomass also giving hydrogen demands very high temperatures (>900 K) (Figure 1). Recently, Zhang et al. [7] and Park et al. [8] demonstrated the proof of concept for such an approach.

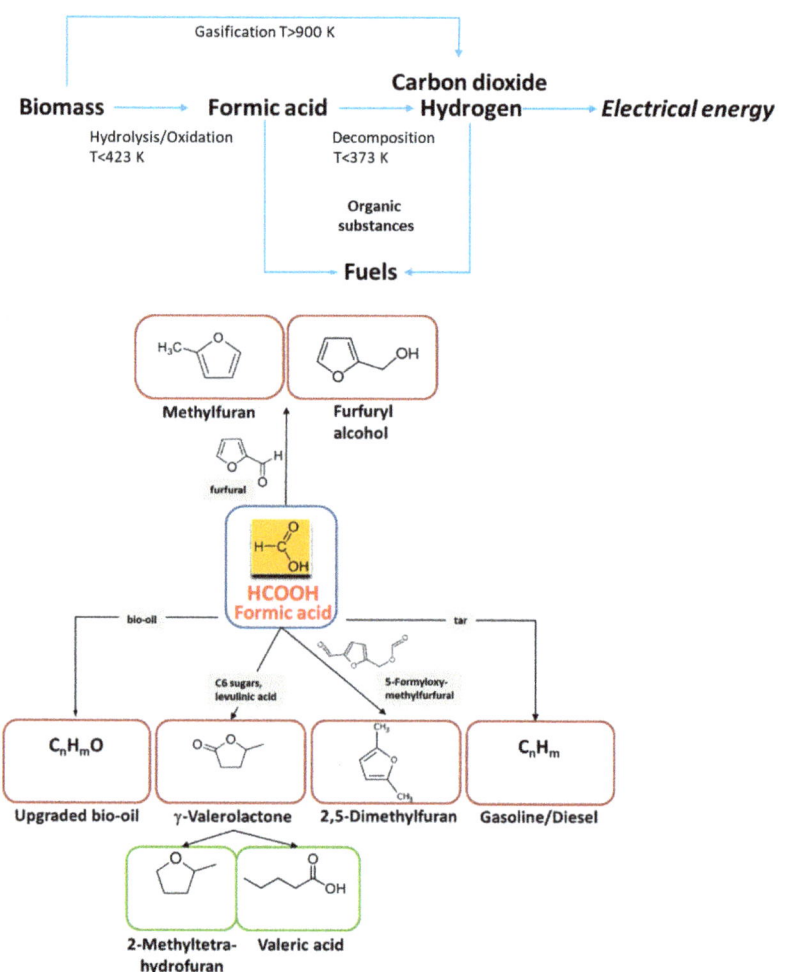

Figure 1. Reaction scheme showing the production of electrical energy and fuels from biomass through the conversion to formic acid.

The hydrogen obtained from formic acid could be further transformed to electrical energy (Figure 1). The development of a compact integrated 25 kW system which converts formic acid to power has been discussed [9]. Formic acid could be used also as a donor of hydrogen instead of molecular hydrogen to hydrogenate different organic substances for production of fuels and intermediates for fuels [4,10]. Thus, it could be applied for synthesis of γ-valerolactone from C6 sugars and levulinic acid [11], 2,5-dimethylfuran from 5-formyloxymethylfurfural [12], furfuryl alcohol [13] and methylfuran [14] from furfural, upgraded bio-oil from bio-oil [15], and diesel/gasoline mixtures from tar [16] (Figure 1).

Supported catalysts with nanoparticles are traditional catalysts for the hydrogen production from formic acid in gas and liquid phase. Novel single atom metal catalysts supported on N-doped carbon may provide a higher activity in the formic acid decomposition than the activity of the catalysts with nanoparticles, but the difference is not so significant [17]. The activity of homogeneous metal complexes is often higher [18–24]. Hence, they could be used at lower temperatures. Metal complexes also represent more uniform active sites as compared to metal nanoparticles. Hence, basing on this knowledge the design of the catalyst could be facilitated. However, there are serious problems of application of homogeneous metal complexes as catalysts for different reactions limiting their industrial applications. They include difficulties in separation of a catalyst from the reaction medium and catalyst's recovery, instability of homogeneous catalytic systems, as well as possible corrosive effects of catalyst solutions on the equipment [25].

Separation of the catalysts with noble metals could be important for production of hydrogen from reaction mixtures containing formic acid and obtained from biomass. To solve this problem, metal complexes could be supported on different supports. Serious efforts have been directed toward the immobilization of homogeneous catalysts on supports. Evidently, their catalytic properties could change due to a change of ligand environment, since after supporting the support surface sites become important ligands for metal atoms. These sites may have no direct analogs in solutions [25]. Their nature affects strongly the energy of interaction of metal complexes and resistance of the catalyst to leaching. Additionally, the support may provide surface sites for formic acid activation leading to its faster conversion.

Carbon dioxide is also produced as a by-product during the decomposition of formic acid; however, it can be further hydrogenated into formate salts at low temperatures [26,27]. Earlier, we have analyzed this reaction taking place on different catalysts, particularly on supported metal complexes [27]. In the present review, we will consider in details the catalytic properties of supported Ru, Ir and Fe complexes in the hydrogen production from formic acid. There are only a few studies performed with supported complexes of other metals (Pd, Rh) in this reaction. We have not found a specialized review related to application of supported metal complexes in the hydrogen production from formic acid. However, this subject is worth to discuss since this type of the catalysts shows excellent activity, selectivity and stability in the reaction and can be easily separated from the reaction mixture.

2. Supported Ruthenium Complexes

Ru complexes are among the most active complexes for the hydrogen production from liquid phase formic acid. The group of Laurenczy contributed significantly to the development of these catalysts [20]. In 2009, they reported the results of immobilization of ruthenium(II)–TPPTS (trisulfonated triphenylphosphine) complex on different supports [28]. Among them, they used an ion exchange resin containing basic trimethylammonium groups. The reaction mixture except of formic acid contained sodium formate (9:1). It is known that addition of sodium formate to formic acid should give a higher activity [19,29,30]. The results showed that this resin ionically interacted strongly with the Ru complex and no Ru leaching took place during the reaction. However, recycling of the catalyst led to a decrease in the reaction rate, but the same conversion was achieved in 3 h.

In another case, covalent interaction of the Ru species to the phosphine groups of PPh_3- or PPh_3-O-cross-linked polystyrene led to strong coordination of the metal. Unfortunately, the obtained catalyst was not sufficiently active in the reaction as compared to the homogeneous catalyst. The authors supposed that the reasons are related to a high hydrophobicity of the material and mass transfer limitations.

Additionally, they used five different types of zeolites as supports. The activity was sufficiently good, but the adsorbed Ru–TPPTS could be removed in water from zeolites thus indicating that it was attached weakly through physical adsorption. This complicates the catalyst recycling which is necessary for a sustainable process.

The same group has noticed later that their earlier attempts to create heterogenized metal complex catalysts were only partially successful and developed another system [29]. In this case, they used a mesoporous silica support (MCM-41) with attached phosphine groups. The optimized catalytic system corresponded to MCM41-Si-$(CH_2)_n$$PPh_2$/Ru-mTPPTS with $n = 2$ and demonstrated the activity and stability comparable to those of the homogeneous catalyst (Figure 2). Thus, the turnover frequency (TOF) of 2780 h^{-1} was obtained at 383 K (Table 1). TOF value corresponds to the number of hydrogen molecules obtained per one metal site per time unit. It is a major value characterizing the specific activity of the catalysts.

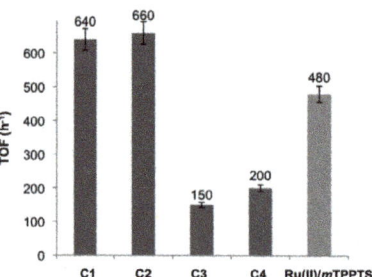

Figure 2. Effect of the number of CH_2 groups attaching phosphine groups to the MCM-41 support on formic acid (HCOOH) decomposition catalyzed by the immobilized Ru catalysts at 363 K. Reprinted with permission from [29].

The effect of the number of CH_2 groups attaching the phosphine groups (n) is demonstrated in Figure 2. It is seen that short CH_2 chains lead to the activity higher than those for the catalysts with longer chains. The content of CO obtained as a by-product was negligible (3 ppm). It is very important to have a very low level of CO in the reaction to prevent poisoning of the catalyst in a fuel cell. In addition, the supported Ru catalysts were recyclable since they allowed performing the reaction for more than 20 cycles without loss in activity.

Later, the same group created a reactor system for continuous production of hydrogen from formic acid [31]. A Ru-mTPPTS catalyst supported on phosphinated polystyrene beads was used in this case. This catalyst provided the TOF of 270 h^{-1} at 378 K and the apparent activation energy of 93.6 kJ mol^{-1}. The low CO concentration level (<5 ppm) was reached due to a PROX reaction using a Pt/CeO_2 catalyst. Alternatively, a methanation of CO could be used to decrease the CO content [9].

Zhao [32] modified the surface of SiO_2 support (450 m^2 g^{-1}) with 3-mercaptopropyltrimethoxysilane to obtain SiO_2-SH groups, which then interacted with Ru or Pd chlorides (about 2 wt %). The obtained Ru-S-SiO_2 and Pd-S-SiO_2 catalysts showed TOFs of 344 h^{-1} and 719 h^{-1} with a 4 M HCOOH/HCOONa (9:1) mixture at 358 K, respectively. X-Ray Photoelectron Spectroscopy (XPS) studies before and after experiments showed the presence of mainly Pd^{2+} ions in the catalyst indicating that they are the active sites of the reaction. The authors also showed that sulfates accelerate the reaction by up to 70%. This could be useful for practical applications.

Wang et al. [33] used a ruthenium pincer complex knitted in a porous organic polymer (810 m^2 g^{-1}). Thermal gravimetric analysis revealed that the supported complex was thermally stable up to 533 K. However, the TOF of 266 h^{-1} obtained at 363 K was not very high. The authors proposed a mechanism for the formic acid decomposition and production based on participation of Ru hydride in the reaction (Figure 3). For the decomposition, the mechanism involves the dissociative formic acid adsorption and CO_2 release followed by H_2 release. It is interesting that N sites of the complex provide deprotonation of the formic acid through dissociation of the O–H bond.

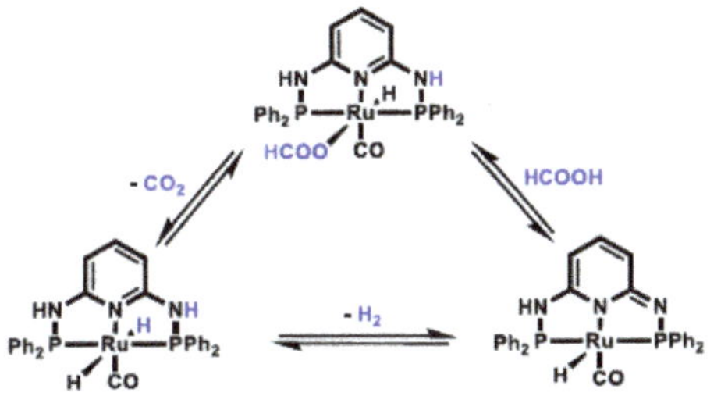

Figure 3. Proposed mechanism for the Ph-PN$_3$P Ru-catalyzed formic acid dehydrogenation and CO$_2$ hydrogenation. Reprinted with permission from [33].

Solakidou et al. [34] showed that amino functionalized silica (H$_2$N@SiO$_2$) significantly increases the TOFs of the hydrogen production from formic acid by a (Ru/P(CH$_2$CH$_2$PPh$_2$)$_3$) homogeneous catalyst. The maximal TOF reached 983 h^{-1} at 363 K. They observed a significant decrease of the apparent activation energy from 41 kJ mol^{-1} to 28 kJ mol^{-1} and supposed that the H$_2$N@SiO$_2$ particles play a dual role: they act as a co-catalyst for deprotonating formic acid by amine groups, and they serve as a template, which stabilizes the metal complex on its surface, thus promoting formate decomposition via (Ru^{II}-hydride) species.

Bavykina et al. [35] used a Ru complex supported on covalent triazine framework (Ru^{II}(η_6-C$_6$H$_6$)/CTF) and obtained high TOF values at 353 K in base free conditions (without Na formate) (4020 h^{-1}, Table 1).

Hausoul et al. [36] studied the effect of the nature of polymeric support on the properties of a Ru complex in the hydrogen production from formic acid. Polymeric analogs of PPh$_3$ (pTPP), 1,2-bis(diphenylphosphino)ethane (pDPPE) (Figure 4), and 1,2-(diphenylphosphino)benzene (pDPPBe) have been tested. The highest TOF of 22,900 h^{-1} was obtained with a RuCl$_2$(p-cymene)/pDPPE catalyst at 433 K (Table 1). The catalyst performed efficiently in solutions with up to 30 wt % formic acid. It is seen in Figure 5a that the activity of the unsupported (RuCl$_2$(p-cymene)(PPh$_3$)) complex is significantly lower than those of the supported complexes. The kinetics of the reaction was featured by an induction period and a pseudo-zero-order dynamics of the pressure increase. Recycling experiments revealed only low leaching and a small decrease in the activity over 7 runs. A Ru/C catalyst with nanoparticles gave a significantly lower activity, lower selectivity and leaching of Ru to the reaction solution.

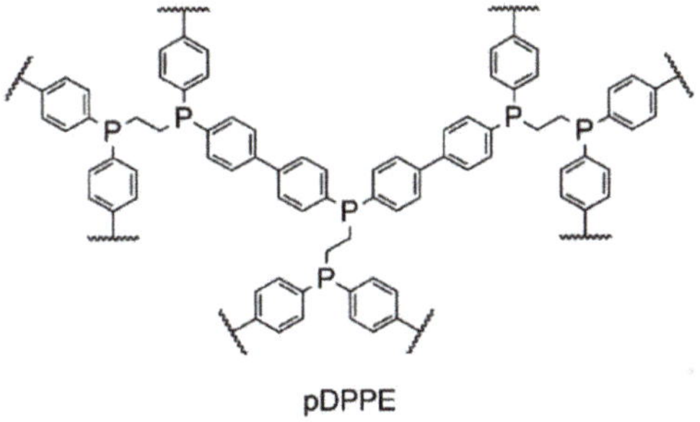

Figure 4. pDPPE.

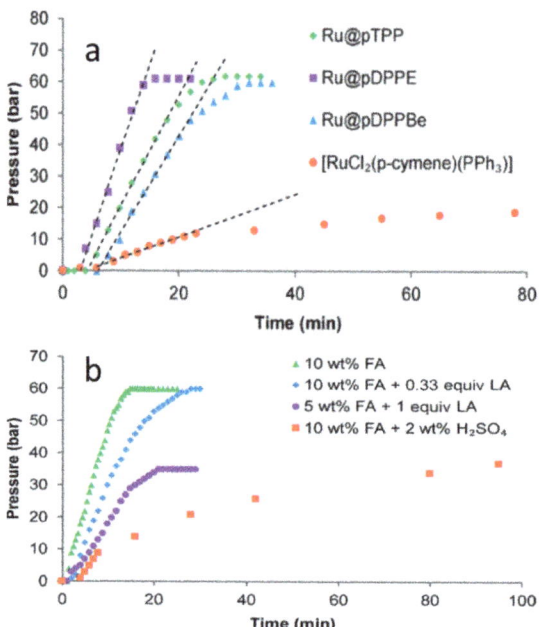

Figure 5. (**a**) Reactor pressure in the Ru-catalyzed decomposition of aqueous formic acid at 433 K. (**b**) Effect of admixtures on decomposition of formic acid on the Ru@pDPPE catalyst. Reprinted with permission from [36].

The same authors studied the decomposition of formic acid in solutions with other substances (Figure 5b). It is seen that levulinic acid (LA) and sulfuric acid retard the reaction, but they do not poison the catalyst completely. This is important to know for development of future biorefineries involving the process of conversion of biomass to hydrogen through the formic acid production [7,8] (Figure 1).

Beloqui Redondo et al. [37] used a 0.4 wt % Ru phosphine complex supported on a metal organic framework (MOF) for the gas phase decomposition of formic acid at 418 K. They obtained 99% selectivity and TOF of 2300 h^{-1} (Table 1). This TOF is sufficiently high for the gas phase decomposition. The authors indicated that phosphine species interact with Ru species providing the formation of Ru single-sites on the MOF support. Amine linkers present in the support could activate formic acid for the reaction through deprotonation. The authors observed an induction period, which was assigned to removal of chloride ligands from the metal complex followed by coordination of formates. However, the Brunauer–Emmett–Teller (BET) surface area of their catalyst decreased significantly after the reaction. It is not clear whether this will take place further and affect negatively the catalytic reaction.

3. Supported Iridium Complexes

As we showed above, several Ru complexes, which are very active in the formic acid decomposition, involve phosphine ligands. Broicher et al. [38] indicated that P-based ligands are sensitive to oxidation, while N-based ligands show a great advantage allowing handling and storage of the catalyst in air. In this section, we will consider Ir complexes, which are mainly attached to N-containing ligands of the support.

Bavykina et al. [35] have used an [IrCp*(OH)](OTf)$_2$ complex (OTf-triflate) to deposit over a covalent triazine framework (CTF) prepared at 773 K with a high surface area (Figure 6a,b). OTf was washed out during the recycling of the catalyst, pointing that formate replaces triflate. The TOFs of the catalysts with a low concentration of the metal complex (0.2 wt %) were higher than those of the catalysts with a high concentration and corresponded to 27,000 h^{-1} (Table 1).

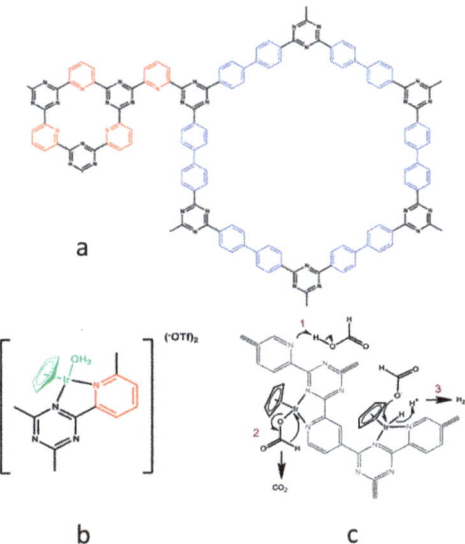

Figure 6. (**a**) Covalent triazine framework (CTF) support unit, (**b**) [IrCp*(OH)](OTf)$_2$ complex attached to the CTF support, and (**c**) scheme showing a catalytic cycle within the CTF support involving the steps of (1) formic acid interaction, (2) β-hydride elimination, and (3) hydrogen release. Adapted with permission from [35].

The authors tested the catalyst's durability in continuous mode. Thus, a highly concentrated formic acid solution (88 wt %) was fed to a reactor at 353 K. At termination of this experiment, a turnover number (TON = the number of H$_2$ molecules related to the number of metal sites) of 1,060,000 was obtained, which demonstrates that the catalyst is highly durable and can be used in devices producing hydrogen.

Figure 6c displays possible molecular pathways for the system [35], which consist of the three main steps: (1) formic acid deprotonation, (2) β-hydride elimination and (3) hydrogen release. The deprotonation is important and takes place on free pyridinic sites that provide basicity of the CTF support. The hydrogen release step has been proposed to be rate-determining.

Gunasekar et al. [39] studied an [IrCp*Cl$_2$]$_2$ complex supported on CTFs prepared at different temperatures 673 and 773 K. The activity of the supported complex was higher than that of the homogeneous complex. The TOFs were, however, lower than those obtained by Bavykina et al. [35] and corresponded to 7930 h^{-1} at 363 K. This could be related to a much higher concentration of metal in the samples. The TOF for the similar supported RhCp* complex was lower than that of the IrCp* complex.

Shen et al. [30] studied an IrCp*Cl$_2$ complex supported on porous polypyrrole particles (500 nm) with a BET surface area of 51 m^2 g^{-1} (Table 1). The TOF was very high in the presence of sodium formate and equal to 46,000 h^{-1} at 363 K. The apparent activation energy in the formic acid decomposition in the absence of sodium formate corresponded to 63 kJ mol^{-1} and in the presence of sodium formate it was approximately the same (66 kJ mol^{-1}). When 1.0 M formic acid solution flowed into a tubular reactor (55 × 9 mm) with a catalyst containing 250 mg of the IrCp*Cl$_2$(polypyrolle) complex at 313 K, the initial hydrogen evolving rate was 5.6 mL min^{-1}, which could generate about 1.1 W electric power through a proton-exchange membrane fuel cell. The authors noted that this value was sufficient to drive, for example, a personal mobile phone.

Recently, Broicher et al. [38] also used an [IrCp*Cl$_2$]$_2$ complex as a precatalyst and a conjugated microporous polymer (CMP) with bipyridine groups as a support. The combination of those gave an Ir@CMP catalyst (Figure 7). This catalyst showed a record value of TOF of 123,894 h^{-1} at 433 K (Table 1), relatively high apparent activation barrier of 90 kJ mol^{-1} and low leaching. The CO content was stable in the range around 68 ppm. Complete conversion of formic acid could be reached.

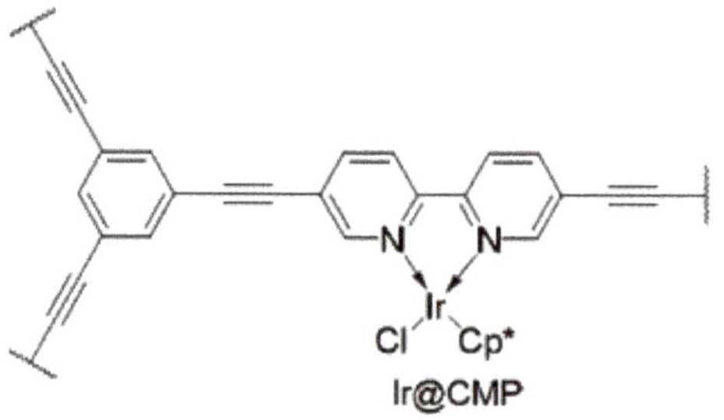

Figure 7. Ir@CMP catalyst.

In the same conditions a commercial Ir/C catalyst with nanoparticles demonstrated a strong leaching and gave a low conversion of formic acid confirming that Ir nanoparticles are not active in the reaction. Application of the [IrCp*Cl$_2$]$_2$ complex without bipyridine ligands as a catalyst gave also only a low conversion. At the same time, the [IrCp*Cl$_2$/2,2-bipy] complex showed a high activity of 43,051 h^{-1} at 433 K demonstrating the need for bipyridine ligands. The activity of the supported 1 wt % Ir@CMP catalyst (Figure 8) was close (35,246 h^{-1}) indicating that the heterogenization affected only weakly. After the reaction, a significant amount of nanoparticles (up to 5 nm) was found in the sample, demonstrating that the measured activity is really caused by a low number of active sites with a very high activity. This point is also supported by a study of the effect of variation of metal loading showing that the TOFs are higher for the catalysts with a low content of metal complex (Figure 8) in accordance with the data of Bavykina et al. [35].

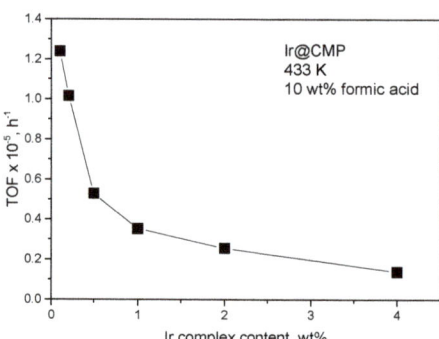

Figure 8. Effect of concentration of IrCp* complex supported on bipyridine-based conjugated microporous polymer on the TOFs in formic acid decomposition. The data are taken with permission from Reference [38].

4. Supported Iron Complexes

The use of efficient non-noble metal catalysts would be a good choice for the hydrogen production from formic acid decomposition. Boddien et al. [40] proposed different homogeneous Fe containing complexes for this reaction. Using a mixture of 0.005 mol% Fe(BF$_4$)$_2$·6H$_2$O and tris[(2-diphenylphosphino)ethyl]phosphine [P(CH$_2$CH$_2$PPh$_2$)$_3$] to a solution of formic acid in propylene carbonate, without other additives or bases, obtained a

TOF value up to 9425 h^{-1} and a TON value of more than 92,000 at 353 K [40]. The apparent activation energy corresponded to 77 kJ mol^{-1} and the CO content did not exceed 20 ppm. However, the catalyst was completely poisoned and became inactive after 16 h of the continuous reaction. This was assigned to chloride and/or water admixtures accumulation on the catalyst.

Later, Stathi et al. [41] successfully heterogenized Fe phosphine complexes on the surface of two types of silica modified with phosphines (Figure 9a). Heterogenization of the Fe(II)/P(CH$_2$CH$_2$PPh$_2$)$_3$ system increased its TOF by 1.7 times as compared to the homogeneous catalyst and was in the range of 6000–8000 h^{-1} (Table 1). The reaction was performed in propylene carbonate solvent. The apparent activation energies were significantly lower than that of the homogeneous complex and corresponded to 51 kJ mol^{-1} and 43 kJ mol^{-1} for Fe^{II}/RPPh$_2$@SiO$_2$ and Fe^{II}/polyRPhphos@SiO$_2$, respectively. The authors indicated that the possible rate determining steps could be hydride elimination or direct hydride transfer from formate to Fe. No leaching of iron in the reaction solution was found. The Fe^{II}/RPPh$_2$@SiO$_2$ catalyst showed a TON of higher than 176,000.

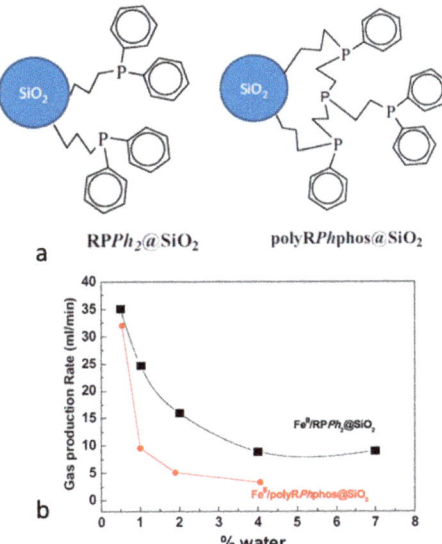

Figure 9. (**a**) Silica supports modified with phosphines. (**b**) Effect of water addition on the formic acid decomposition over supported Fe containing catalysts at 373 K. Reprinted with permission from [41].

The authors observed an inhibitory water effect (Figure 9b), but it was reversible, since the catalyst could be reactivated by a simple washing. This result is important to know, since formic acid always contains a small amount of water, which can be formed together with carbon monoxide due to self-decomposition of formic acid taking place during its storage [42]. This effect is more significant at high concentrations of formic acid.

Table 1. Properties of the most active supported Ru, Ir and Fe complexes used for the hydrogen production from formic acid.

Initial or Attached Complex	Catalyst Support	BET Surface Area of the Support, m² g⁻¹	Active Metal Concentration, wt %	T, K	Concentration of Formic Acid and Sodium Formate	TOF, h⁻¹ (Ea, kJ mol⁻¹)	Reference
Ru-mTPPTS	MCM41-Si(CH₂)₂PPh₂	-	0.3	383	10 M (HCOOH+HCOONa, 9:1)	2780	[28]
Ru^II(η₆-C₆H₆)	CTF500	1800	2.7	353	3 M	4020	[35]
RuCl₂(p-cymene)	pDPPE	33	1	433	2.2 M	22,900	[36]
RuCl₂(p-cymene)	PPh₂-MOF	1075 to 161 (after reaction)	0.7	418	5 vol% (gas phase reaction)	2300	[37]
Ir^III Cp*	CTF500	1800	0.2	353	3 M	27,000	[35]
[Cp*IrCl₂]₂	bpy-CTF400	684	1.4	353	1 M	2820	[39]
[Cp*IrCl₂]₂	bpy-CTF500	1566	11.3	363	1 M	7930	[39]
Cp*IrCl₂	polypyrrole	51	4.3	333	1 M	4060	[30]
Cp*IrCl₂	polypyrrole	51	4.3	363	2 M (HCOOH+HCOONa,1:1)	46,000 (66)	[30]
[Cp*IrCl₂]₂	CMP	706	0.1	433	2.2 M	123,894 (90)	[38]
Fe(BF₄)₂	polyRPhphos@SiO₂	502	0.8	363	7.6 M	7600 (51)	[41]
Fe(BF₄)₂	RPPh₂@SiO₂	531	0.9	363	7.6 M	6396 (43)	[41]

Ea—apparent activation energy, mTPPTS—meta-trisulfonated triphenylphosphine, PPh—phenylphosphine, CTF—covalent triazine framework, MOF—metal organic framework, Cp*—pentamethylcyclopentadienyl, CMP—bipyridine-based conjugated microporous polymer.

5. Discussion

The summarized data for the key catalysts with supported Ru, Ir and Fe complexes for the hydrogen production from formic acid are shown in Table 1. Other metal complexes are almost not studied. The table can help to choose the optimal catalysts corresponding to certain conditions of the reaction. However, it is not easy to compare the activity of the catalysts (TOFs) presented in Table 1, since the conditions of the reaction and concentrations of the active component in the catalysts were different. Moreover, some experiments have been performed in the presence of sodium formate. Basic additives to the reaction mixture or basic sites of the catalysts/supports are known to promote significantly the reaction. They deprotonate formic acid to formate species. Deprotonation of formic acid can be provided also by traditional oxide supports having basic sites [43] and by introduction of alkali metals promoters to supported metal catalysts [44–46]. Deprotonation provided by pyridinic N sites of N-doped carbon support was also reported for the catalysts with single metal atoms [47,48]. In this case, it reminds the effect of basic amine additives.

As for catalysts with nanoparticles, the steady-state TOF values obtained for the gas-phase reaction over a Pd/C catalyst doped with K ions with 3–4 nm Pd nanoparticles did not exceed 3600 h^{-1} at 353 K [44,45]. For the liquid-phase reaction and Pd nanoparticles (~1.4 nm) supported on N-doped carbon, the initial values were higher and reached 8414 h^{-1} at 333 K in the presence of sodium formate [22]. Some supported catalysts with nanoparticles (Ir/C [38] and Ru/C [36]) were used for comparison of the activity with supported metal complexes (Table 1). It was shown that their activity is negligible as compared with the activity of supported metal complexes. The disadvantage of these comparisons was that the mean sizes of nanoparticles in the catalysts with nanoparticles were not reported.

In contrast, some homogeneous complexes showed much higher TOFs than those of the supported metal complexes (Table 1). The values in the range 250,000–322,000 h^{-1} for temperatures 363 and 373 K have been reported by a few groups of authors [21,23,24]. These complexes are also based on Ir [23,24] and Ru [21]. It would be useful to immobilize them on some supports in order to have an opportunity to separate easily the obtained catalyst from the reaction mixture.

As it is shown above (Figure 8), concentration of a metal complex is an important factor determining TOFs. Interesting that at a lower concentration of a metal complex higher TOFs were observed. In this case, the active sites could be stabilized by specific support sites. The nature of these active sites should be studied using advanced methods like extended X-ray absorption fine structure (EXAFS) combined with X-ray absorption near edge structure (XANES) preferably in situ. Using density functional theory (DFT) calculations may assist in understanding the structure of these active sites. There is an evident lack of such studies. The progress in understanding may lead to development of a targeted synthesis of the catalysts with these very active sites.

6. Conclusions

Therefore, the analysis of the literature showed that immobilization of Ru, Ir and Fe complexes on some polymers, covalent triazine frameworks, metal organic frameworks or silica modified with phosphines is promising for the hydrogen production from formic acid in terms of activity of the catalysts and possibility to separate the catalysts from the reaction mixture. Supported Ir complexes were normally more efficient than the supported Ru complexes. Thus, the maximal TOF value was reached for the IrCp* complex supported on the bipyridine-based conjugated microporous polymer and corresponded to 123,894 h^{-1} at 433 K. In part, the high activity of Ir complexes could be provided by the presence of basic N sites of the supports which deprotonate formic acid for further easier decomposition of the formed formate species with participation of the Ir atoms. Yet, the activities of some homogeneous metal complexes were higher and reached 322,000 h^{-1}.

A strong concentration effect of metal complexes was observed demonstrating that at a small concentration of supported metal complex higher TOFs are obtained than those at a high concentration. This effect is not related to nanoparticles formation. Finally, non-noble metal supported Fe complexes were efficient in the reaction provided the contents of water and chlorine ions in the solution were negligible.

Funding: The work was funded by the Russian Science Foundation (grant 17-73-30032).

Institutional Review Board Statement: Not applicable.

Informed Consent Statement: Not applicable.

Data Availability Statement: Data is contained within the article.

Conflicts of Interest: The authors declare no conflict of interest.

References

1. IEA. The Future of Hydrogen. Available online: www.iea.org/reports/the-future-of-hydrogen (accessed on 26 February 2021).
2. Preuster, P.; Papp, C.; Wasserscheid, P. Liquid Organic Hydrogen Carriers (LOHCs): Toward a Hydrogen-free Hydrogen Economy. *Acc. Chem. Res.* **2017**, *50*, 74–85. [CrossRef] [PubMed]
3. Rao, P.C.; Yoon, M. Potential Liquid-Organic Hydrogen Carrier (LOHC) Systems: A Review on Recent Progress. *Energies* **2020**, *13*, 6040. [CrossRef]
4. Bulushev, D.A.; Ross, J.R.H. Towards Sustainable Production of Formic Acid. *ChemSusChem* **2018**, *11*, 821–836. [CrossRef]
5. Gromov, N.V.; Medvedeva, T.B.; Rodikova, Y.A.; Babushkin, D.E.; Panchenko, V.N.; Timofeeva, M.N.; Zhizhina, E.G.; Taran, O.P.; Parmon, V.N. One-pot synthesis of formic acid via hydrolysis–oxidation of potato starch in the presence of cesium salts of heteropoly acid catalysts. *RSC Adv.* **2020**, *10*, 28856–28864. [CrossRef]
6. Preuster, P.; Albert, J. Biogenic Formic Acid as a Green Hydrogen Carrier. *Energy Technol.* **2018**, *6*, 501–509. [CrossRef]
7. Zhang, P.; Guo, Y.-J.; Chen, J.; Zhao, Y.-R.; Chang, J.; Junge, H.; Beller, M.; Li, Y. Streamlined hydrogen production from biomass. *Nat. Catal.* **2018**, *1*, 332–338. [CrossRef]
8. Park, J.-H.; Lee, D.-W.; Jin, M.-H.; Lee, Y.-J.; Song, G.-S.; Park, S.-J.; Jung, H.J.; Oh, K.K.; Choi, Y.-C. Biomass-formic acid-hydrogen conversion process with improved sustainability and formic acid yield: Combination of citric acid and mechanocatalytic depolymerization. *Chem. Eng. J.* **2020**, 127827. [CrossRef]
9. Van Putten, R.; Wissink, T.; Swinkels, T.; Pidko, E.A. Fuelling the hydrogen economy: Scale-up of an integrated formic acid-to-power system. *Int. J. Hydrogen Energy* **2019**, *44*, 28533–28541. [CrossRef]
10. Nie, R.; Tao, Y.; Nie, Y.; Lu, T.; Wang, J.; Zhang, Y.; Lu, X.; Xu, C.C. Recent Advances in Catalytic Transfer Hydrogenation with Formic Acid over Heterogeneous Transition Metal Catalysts. *ACS Catal.* **2021**, *11*, 1071–1095. [CrossRef]
11. Heeres, H.; Handana, R.; Chunai, D.; Borromeus Rasrendra, C.; Girisuta, B.; Jan Heeres, H. Combined dehydration/(transfer)-hydrogenation of C6-sugars (D-glucose and D-fructose) to γ-valerolactone using ruthenium catalysts. *Green Chem.* **2009**, *11*, 1247–1255. [CrossRef]
12. Sun, Y.; Xiong, C.; Liu, Q.; Zhang, J.; Tang, X.; Zeng, X.; Liu, S.; Lin, L. Catalytic Transfer Hydrogenolysis/Hydrogenation of Biomass-Derived 5-Formyloxymethylfurfural to 2,5-Dimethylfuran Over Ni–Cu Bimetallic Catalyst with Formic Acid As a Hydrogen Donor. *Ind. Eng. Chem. Res.* **2019**, *58*, 5414–5422. [CrossRef]
13. Nagaiah, P.; Gidyonu, P.; Ashokraju, M.; Rao, M.V.; Challa, P.; Burri, D.R.; Kamaraju, S.R.R. Magnesium Aluminate Supported Cu Catalyst for Selective Transfer Hydrogenation of Biomass Derived Furfural to Furfuryl Alcohol with Formic Acid as Hydrogen Donor. *ChemistrySelect* **2019**, *4*, 145–151. [CrossRef]
14. Fu, Z.; Wang, Z.; Lin, W.; Song, W.; Li, S. High efficient conversion of furfural to 2-methylfuran over Ni-Cu/Al$_2$O$_3$ catalyst with formic acid as a hydrogen donor. *Appl. Catal. A Gen.* **2017**, *547*, 248–255. [CrossRef]
15. Bulushev, D.A.; Ross, J.R.H. Catalysis for conversion of biomass to fuels via pyrolysis and gasification: A review. *Catal. Today* **2011**, *171*, 1–13. [CrossRef]
16. Chesnokov, V.V.; Dik, P.P.; Chichkan, A.S. Formic Acid as a Hydrogen Donor for Catalytic Transformations of Tar. *Energies* **2020**, *13*, 4515. [CrossRef]
17. Bulushev, D.A.; Bulusheva, L.G. Catalysts with single metal atoms for the hydrogen production from formic acid. *Catal. Rev.* **2021**, 1–40. [CrossRef]
18. Li, Z.; Xu, Q. Metal-Nanoparticle-Catalyzed Hydrogen Generation from Formic Acid. *Acc. Chem. Res.* **2017**, *50*, 1449–1458. [CrossRef]
19. Stathi, P.; Solakidou, M.; Louloudi, M.; Deligiannakis, Y. From Homogeneous to Heterogenized Molecular Catalysts for H$_2$ Production by Formic Acid Dehydrogenation: Mechanistic Aspects, Role of Additives, and Co-Catalysts. *Energies* **2020**, *13*, 733. [CrossRef]
20. Laurenczy, G.; Dyson, P.J. Homogeneous Catalytic Dehydrogenation of Formic Acid: Progress Towards a Hydrogen-Based Economy. *J. Braz. Chem. Soc.* **2014**, *25*, 2157–2163. [CrossRef]
21. Filonenko, G.A.; van Putten, R.; Schulpen, E.N.; Hensen, E.J.M.; Pidko, E.A. Highly Efficient Reversible Hydrogenation of Carbon Dioxide to Formates Using a Ruthenium PNP-Pincer Catalyst. *ChemCatChem* **2014**, *6*, 1526–1530. [CrossRef]

22. Onishi, N.; Iguchi, M.; Yang, X.; Kanega, R.; Kawanami, H.; Xu, Q.; Himeda, Y. Development of Effective Catalysts for Hydrogen Storage Technology Using Formic Acid. *Adv. Energy Mater.* **2019**, *9*, 1801275. [CrossRef]
23. Papp, G.; Ölveti, G.; Horváth, H.; Kathó, Á.; Joó, F. Highly efficient dehydrogenation of formic acid in aqueous solution catalysed by an easily available water-soluble iridium(iii) dihydride. *Dalton Trans.* **2016**, *45*, 14516–14519. [CrossRef] [PubMed]
24. Wang, W.-H.; Ertem, M.Z.; Xu, S.; Onishi, N.; Manaka, Y.; Suna, Y.; Kambayashi, H.; Muckerman, J.T.; Fujita, E.; Himeda, Y. Highly Robust Hydrogen Generation by Bioinspired Ir Complexes for Dehydrogenation of Formic Acid in Water: Experimental and Theoretical Mechanistic Investigations at Different pH. *ACS Catal.* **2015**, *5*, 5496–5504. [CrossRef]
25. Yermakov, Y.I.; Kuznetsov, B.N.; Zakharov, V.A. Chapter 1: Introduction to the Field of Catalysis by Supported Complexes. In *Studies in Surface Science and Catalysis*; Elsevier: Amsterdam, The Netherlands, 1981; Volume 8, pp. 1–58. [CrossRef]
26. Álvarez, A.; Bansode, A.; Urakawa, A.; Bavykina, A.V.; Wezendonk, T.A.; Makkee, M.; Gascon, J.; Kapteijn, F. Challenges in the Greener Production of Formates/Formic Acid, Methanol, and DME by Heterogeneously Catalyzed CO_2 Hydrogenation Processes. *Chem. Rev.* **2017**, *117*, 9804–9838. [CrossRef]
27. Bulushev, D.A.; Ross, J.R.H. Heterogeneous catalysts for hydrogenation of CO_2 and bicarbonates to formic acid and formates. *Catal. Rev.* **2018**, *60*, 566–593. [CrossRef]
28. Gan, W.; Dyson, P.J.; Laurenczy, G. Hydrogen storage and delivery: Immobilization of a highly active homogeneous catalyst for the decomposition of formic acid to hydrogen and carbon dioxide. *React. Kinet. Catal. Lett.* **2009**, *98*, 205. [CrossRef]
29. Gan, W.; Dyson, P.J.; Laurenczy, G. Heterogeneous Silica-Supported Ruthenium Phosphine Catalysts for Selective Formic Acid Decomposition. *ChemCatChem* **2013**, *5*, 3124–3130. [CrossRef]
30. Shen, Y.; Zhan, Y.; Bai, C.; Ning, F.; Wang, H.; Wei, J.; Lv, G.; Zhou, X. Immobilized iridium complexes for hydrogen evolution from formic acid dehydrogenation. *Sustain. Energy Fuels* **2020**, *4*, 2519–2526. [CrossRef]
31. Yuranov, I.; Autissier, N.; Sordakis, K.; Dalebrook, A.F.; Grasemann, M.; Orava, V.; Cendula, P.; Gubler, L.; Laurenczy, G. Heterogeneous Catalytic Reactor for Hydrogen Production from Formic Acid and Its Use in Polymer Electrolyte Fuel Cells. *ACS Sustain. Chem. Eng.* **2018**, *6*, 6635–6643. [CrossRef]
32. Zhao, Y.; Deng, L.; Tang, S.-Y.; Lai, D.-M.; Liao, B.; Fu, Y.; Guo, Q.-X. Selective Decomposition of Formic Acid over Immobilized Catalysts. *Energy Fuels* **2011**, *25*, 3693–3697. [CrossRef]
33. Wang, X.; Ling, E.A.P.; Guan, C.; Zhang, Q.; Wu, W.; Liu, P.; Zheng, N.; Zhang, D.; Lopatin, S.; Lai, Z.; et al. Single-Site Ruthenium Pincer Complex Knitted into Porous Organic Polymers for Dehydrogenation of Formic Acid. *ChemSusChem* **2018**, *11*, 3591–3598. [CrossRef]
34. Solakidou, M.; Deligiannakis, Y.; Louloudi, M. Heterogeneous amino-functionalized particles boost hydrogen production from Formic Acid by a ruthenium complex. *Int. J. Hydrogen Energy* **2018**, *43*, 21386–21397. [CrossRef]
35. Bavykina, A.V.; Goesten, M.G.; Kapteijn, F.; Makkee, M.; Gascon, J. Efficient production of hydrogen from formic acid using a Covalent Triazine Framework supported molecular catalyst. *ChemSusChem* **2015**, *8*, 809–812. [CrossRef]
36. Hausoul, P.J.C.; Broicher, C.; Vegliante, R.; Göb, C.; Palkovits, R. Solid Molecular Phosphine Catalysts for Formic Acid Decomposition in the Biorefinery. *Angew. Chem. Int. Ed.* **2016**, *55*, 5597–5601. [CrossRef]
37. Beloqui Redondo, A.; Morel, F.L.; Ranocchiari, M.; van Bokhoven, J.A. Functionalized Ruthenium–Phosphine Metal–Organic Framework for Continuous Vapor-Phase Dehydrogenation of Formic Acid. *ACS Catal.* **2015**, *5*, 7099–7103. [CrossRef]
38. Broicher, C.; Foit, S.R.; Rose, M.; Hausoul, P.J.C.; Palkovits, R. A Bipyridine-Based Conjugated Microporous Polymer for the Ir-Catalyzed Dehydrogenation of Formic Acid. *ACS Catal.* **2017**, *7*, 8413–8419. [CrossRef]
39. Gunasekar, G.H.; Kim, H.; Yoon, S. Dehydrogenation of formic acid using molecular Rh and Ir catalysts immobilized on bipyridine-based covalent triazine frameworks. *Sustain. Energy Fuels* **2019**, *3*, 1042–1047. [CrossRef]
40. Boddien, A.; Mellmann, D.; Gärtner, F.; Jackstell, R.; Junge, H.; Dyson, P.J.; Laurenczy, G.; Ludwig, R.; Beller, M. Efficient Dehydrogenation of Formic Acid Using an Iron Catalyst. *Science* **2011**, *333*, 1733. [CrossRef] [PubMed]
41. Stathi, P.; Deligiannakis, Y.; Avgouropoulos, G.; Louloudi, M. Efficient H_2 production from formic acid by a supported iron catalyst on silica. *Appl. Catal. A Gen.* **2015**, *498*, 176–184. [CrossRef]
42. Hietala, J.; Vuori, A.; Johnsson, P.; Pollari, I.; Reutemann, W.; Kieczka, H. Formic Acid. In *Ullmann's Encyclopedia of Industrial Chemistry*; Wiley-VCH Verlag GmbH & Co. KGaA: Weinheim, Germany, 2016; pp. 1–22. [CrossRef]
43. Zacharska, M.; Chuvilin, A.L.; Kriventsov, V.V.; Beloshapkin, S.; Estrada, M.; Simakov, A.; Bulushev, D.A. Support effect for nanosized Au catalysts in hydrogen production from formic acid decomposition. *Catal. Sci. Technol.* **2016**, *6*, 6853–6860. [CrossRef]
44. Jia, L.; Bulushev, D.A.; Beloshapkin, S.; Ross, J.R.H. Hydrogen production from formic acid vapour over a Pd/C catalyst promoted by potassium salts: Evidence for participation of buffer-like solution in the pores of the catalyst. *Appl. Catal. B-Environ.* **2014**, *160*, 35–43. [CrossRef]
45. Jia, L.; Bulushev, D.A.; Ross, J.R.H. Formic acid decomposition over palladium based catalysts doped by potassium carbonate. *Catal. Today* **2016**, *259*, 453–459. [CrossRef]
46. Bulushev, D.A.; Zacharska, M.; Guo, Y.; Beloshapkin, S.; Simakov, A. CO-free hydrogen production from decomposition of formic acid over Au/Al_2O_3 catalysts doped with potassium ions. *Catal. Commun.* **2017**, *92*, 86–89. [CrossRef]

47. Bing, Q.M.; Liu, W.; Yi, W.C.; Liu, J.Y. Ni anchored C_2N monolayers as low-cost and efficient catalysts for hydrogen production from formic acid. *J. Power Sources* **2019**, *413*, 399–407. [CrossRef]
48. Bulushev, D.A.; Sobolev, V.I.; Pirutko, L.V.; Starostina, A.V.; Asanov, I.P.; Modin, E.; Chuvilin, A.L.; Gupta, N.; Okotrub, A.V.; Bulusheva, L.G. Hydrogen Production from Formic Acid over Au Catalysts Supported on Carbon: Comparison with Au Catalysts Supported on SiO_2 and Al_2O_3. *Catalysts* **2019**, *9*, 376. [CrossRef]

Review

Hydrolytic Dehydrogenation of Ammonia Borane Attained by Ru-Based Catalysts: An Auspicious Option to Produce Hydrogen from a Solid Hydrogen Carrier Molecule

Miriam Navlani-García [1,*], David Salinas-Torres [2] and Diego Cazorla-Amorós [1]

1. Department of Inorganic Chemistry and Materials Institute, University of Alicante, 03080 Alicante, Spain; cazorla@ua.es
2. Department of Physical Chemistry and Materials Institute, University of Alicante, 03080 Alicante, Spain; david.salinas@ua.es
* Correspondence: miriam.navlani@ua.es; Tel.: +34-965-903-400 (ext. 9150)

Abstract: Chemical hydrogen storage stands as a promising option to conventional storage methods. There are numerous hydrogen carrier molecules that afford satisfactory hydrogen capacity. Among them, ammonia borane has attracted great interest due to its high hydrogen capacity. Great efforts have been devoted to design and develop suitable catalysts to boost the production of hydrogen from ammonia borane, which is preferably attained by Ru catalysts. The present review summarizes some of the recent Ru-based heterogeneous catalysts applied in the hydrolytic dehydrogenation of ammonia borane, paying particular attention to those supported on carbon materials and oxides.

Keywords: ammonia borane; hydrogen production; hydrogen carrier; hydrogen storage; Ru nanoparticles

1. Introduction

Energy demand has constantly increased in the last decades, which is closely linked to the expanding population and increasing prosperity. Currently, about 80% of the world's energy supply comes from fossil fuels (i.e., coal, oil, and natural gas). However, their utilization is inevitably associated with the emission of hazardous gases, which drives the current global warming crisis. Currently, nearly 100% of the total CO_2 emissions originate from the combustion and processing of fossil fuels [1], and its concentration in the atmosphere, which has experienced a great increase since the start of the Industrial Revolution, is nowadays higher than 400 ppm [2].

Hence, the tremendous concerns about the environmental issues related to the use of fossil fuels, together with their finite nature, is fostering the search toward the deployment of renewable energy sources. Such an ambitious goal has been the central focus of many investigations. In this sense, the use of hydrogen has tremendous hope for the use of renewable energies in different industrial applications and the transport sector. This is reflected in the constantly increasing number of publications in which its supremacy as an outstanding energy vector is highlighted [3–7].

The use of hydrogen, produced from renewable energy sources, has significant benefits. Its use has zero emission of greenhouse gases and it produces only water as a by-product. Additionally, it has a high energy storage capacity on a gravimetric basis (120 MJ/kg), which is much greater than those of gasoline (44.4 MJ/kg), diesel (45.4 MJ/kg), biodiesel oil (42.2 MJ/kg), and natural gas (53.6 MJ/kg) [8]. However, against all the benefits of hydrogen, its low volumetric energy density (i.e., 0.01 MJ/L at standard temperature and pressure conditions (STP)), which is much lower than those of common fuels (gasoline (34.2 MJ/L), diesel (34.6 MJ/L), biodiesel oil (33 MJ/L), and natural gas (0.0364 MJ/L)) limits its utilization as a fuel at ambient conditions. There are different options to increase the hydrogen energy density so that it can meet the target set by the U.S. Department of Energy (DOE), which fixes the ultimate onboard hydrogen storage for light-duty fuel cell vehicles at 0.065 kg H$_2$/kg system and 0.050 kg/L, in gravimetric and volumetric basis, respectively [9]. The most used options encompass physical methods such as hydrogen liquefaction, compression, and adsorption in porous materials [10,11]. However, the harsh conditions (such as very high pressure or extremely low temperature), and the high cost associated with the infrastructure needed for the safe handling and storage of hydrogen are important drawbacks of such physical storage methods.

In contrast, chemical storage methods stand up as a promising option, which is particularly important for onboard automotive applications, in which the technologies used in the physical methods do not fully meet the DOE targets for safe and inexpensive hydrogen storage [12,13].

Chemical hydrogen storage refers to those processes in which molecular hydrogen is released through a chemical reaction that starts when the hydrogen-containing molecule (i.e., hydrogen carrier) is subjected to thermal or catalytic decomposition [14].

There are various hydrogen carrier molecules in both the liquid and solid phase. Notable examples of those molecules, together with their hydrogen content (in wt. %) are included in Table 1.

Table 1. Examples of hydrogen storage molecules.

Hydrogen Storage Material	State	Hydrogen Content in wt. %	Reference
NH$_3$BH$_3$	Solid	19.5	[15,16]
LiBH$_4$	Solid	18.4	[17,18]
NaBH$_4$	Solid	10.8	[19,20]
MgH$_2$	Solid	7.6	[21,22]
NH$_3$	Liquid	17.6	[23,24]
CH$_3$OH	Liquid	12.6	[25,26]
H$_2$NNH$_2$	Liquid	12.5	[27,28]
H$_2$O	Liquid	11.1	[29,30]
HCOOH	Liquid	4.4	[31–33]

Seeking and exploring new hydrogen storage options are in continuous progress and significant advances related to reversible hydrogen storage have been recently achieved [34–37]. In this review, we cover some of the most relevant recent strategies on hydrogen production from NH$_3$BH$_3$ (ammonia borane, AB), which is the simplest nitrogen boron hydrogen compound [38–40], and one of the most fruitfully investigated solid-state hydrogen-rich molecules.

AB is a white crystalline solid at room temperature, which was first prepared by Shore and Parry in 1955 [41], and it has received tremendous attention due to several advantages compared to other hydrogen carrier molecules such as its high hydrogen content (19.6 wt. %; each equivalent yielding up to 3 equivalents of hydrogen), low molecular weight (30.87 g mol^{-1}), its stability in solid-state, and high solubility in water. Furthermore, it is nonexplosive and non-flammable under standard conditions. AB has a high melting point of 112 °C and a density of 0.74 g cm^{-3} [38].

AB has an equal number of protic H$^{\delta+}$ (N-H) and hydridic H$^{\delta-}$ (B-H) hydrogens in intra- and intermolecular interactions. It has heteropolar (N-H···H-B) and homopolar (B-H···H-B) dihydrogen interactions, which are the origin of the intra- and intermolecular dehydrogenation of AB [42]. It also has a strong B-N bond, so that the release of hydrogen is more favored than the dissociation into NH$_3$ and BH$_3$ under most conditions [43]. AB can be synthesized in three different ways (i.e., Lewis acid–Lewis base exchange, salt-metathesis followed by hydrogen release, and isomerization of the diammonate of diborane ([H$_2$B(NH$_3$)$_2$]$^+$[BH$_4$]$^-$)) [38].

The dehydrogenation of AB can be performed by either thermolysis, a process which needs much thermal energy (it requires temperatures as high as 200 °C), or solvolysis in protic solvents (i.e., hydrolysis in water (Equation (1)) and methanolysis in methanol (Equation (2)) and dehydrocoupling in nonprotic solvents (Equations (3) and (4)) [44].

$$NH_3BH_3 \text{ (aq)} + 2H_2O \text{ (l)} \rightarrow NH_4 \cdot BO_2 \text{ (aq)} + 3H_2 \text{ (g)} \quad (1)$$

$$NH_3BH_3 \text{ (sol)} + 4CH_3OH \text{ (l)} \rightarrow NH_4 \cdot B(OCH_3)_4 \text{ (sol)} + 3H_2 \text{ (g)} \quad (2)$$

$$nNH_3BH_3 \text{ (sol)} \rightarrow (NH_2BH_2)_n \text{ (s or sol)} + nH_2 \text{ (g)} \quad (3)$$

$$nNH_3BH_3 \text{ (sol)} \rightarrow (NHBH)_n \text{ (s or sol)} + nH_2 \text{ (g)} \quad (4)$$

AB solvolysis can afford 3 equivalents of molecular hydrogen at moderate temperatures upon utilization of a proper catalyst, so this is the preferred option. The dehydrogenation of amine-borane adducts catalyzed by transition-metals dates back to the late 1980s [45], but its application in the production of hydrogen is still drawing great interest in the research community, as can be seen in the increasing number of publications reported per year (see Figure 1).

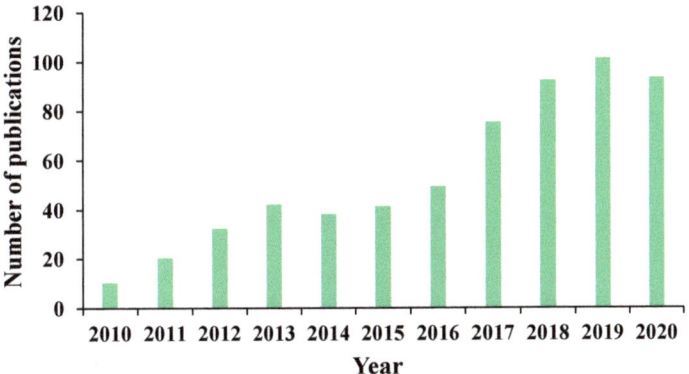

Figure 1. Number of publications in the last 10 years found on the ISI Web of Science for the entry "hydrogen production and ammonia borane".

Both homogeneous and heterogeneous catalytic systems have been explored, but the advantages of heterogeneous catalysts make these systems preferred from a practical point of view. There is vast literature reported on the hydrolytic dehydrogenation of AB, highlighting those contributions made by Yamashita et al. [46–54], Özkar et al. [55–58], and Xu et al. [44,59–62].

Among the heterogeneous catalysts, systems based on ruthenium nanoparticles (NPs) have shown outstanding performances. Ru-based catalysts usually achieve complete AB dehydrogenation, producing ~3 equivalents of hydrogen in short reaction times, and with thermo-controllable reaction rates. The hydrolysis rate is also frequently related to the amount of catalyst and AB, but the hydrolysis rate is frequently found as zero-order relation or quasi-zero-order relation with the concentration of AB [63].

It is worth mentioning that despite the intense efforts devoted to unveiling the mechanism of the hydrolytic dehydrogenation of ammonia borane, there are still some aspects that remain unclear (e.g., rate-determining steps, the order of bond cleavages, etc.) [64]. For instance, Xu et al. postulated that the interaction between AB molecules and the surface of the metal active phase gives rise to the formation of activated complex species, which are attacked by a molecule of H_2O, leading to the concerted dissociation of the B-N bond and the hydrolysis of the BH_3 intermediate to form BO_2, releasing H_2 [65]. Fu et al. proposed a mechanism that proceeds *via* an almost self-powered process that involves the formation of BH_3OH^- and NH_4^+, followed by the attack of adjacent H_2O to generate H_2 [66]. Na et al. suggested that the mechanism is very similar to that of the hydrolytic dehydrogenation of sodium borohydride, and proceeds *via* dissociative adsorption of ammonia borane on Ru surface [67]. More recently, Liu et al. claimed that the hydrogen production from AB attained by noble metal catalysts occurs *via* the following steps: (1) AB molecules interact with the surface of the catalyst to form a complex; (2) A molecule of H_2O attacks to AB-catalyst complex; and (3) AB and H_2O each lose a hydrogen atom to form H_2. Such a mechanism is illustrated in Figure 2 [63].

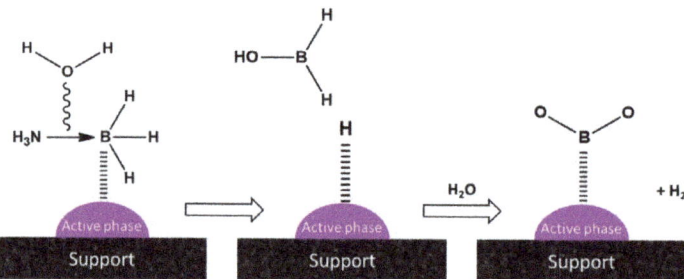

Figure 2. Mechanism proposed for hydrogen production from the hydrolysis of ammonia borane (AB). Adapted from [63].

It is evident that the catalyst's nature plays a crucial role in controlling the whole reaction. Most of the studies reported on Ru-based catalysts for the hydrolytic dehydrogenation of AB are focused on elucidating the role of the features of the metal active phase (i.e., size, morphology, incorporation of a second and third metal in the nanoparticles, etc.), while less attention has been paid to the properties of the support. We divided this manuscript into several sections, which contain a review of representative catalytic systems based on monometallic Ru NPs and supports of a diverse nature, which have been used for the hydrolytic dehydrogenation of AB. As a summary, Table 2 includes representative examples of Ru-based catalysts supported on carbon materials, oxides, metal organic frameworks (MOF), and some other less explored supports, together with the turnover frequency values achieved (TOF; in $mol_{H_2} \cdot mol_{Ru}^{-1} \cdot min^{-1}$) and the calculated activation energy (Ea; in kJ mol^{-1}).

Table 2. Catalytic activity of heterogeneous Ru-based catalysts used for the hydrolytic dehydrogenation of ammonia borane (AB).

Catalyst	TOF ($mol_{H_2} \cdot mol_{Ru}^{-1} \cdot min^{-1}$)	Ea (kJ mol^{-1})	Reference
Ru/Graphene	100	11.7	[68]
Ru/NC-Fe	102.9	47.42	[69]
Ru/nanodiamond	229	50.7	[70]
CF-BT-Ru	322	32.41	[71]
Ru(0)@MWCNT	329	33	[72]
Ru/BC-hs	354	45.72	[73]
Ru/PPC	413	35.2	[74]
Ru/C	429.5	34.81	[75]
Ru/HPCM	440	43.0	[76]
Ru/g-C_3N_4	459.3	37.4	[77]
Ru/Graphene	600	12.7	[78]
Ru/C(800)	670	14.3	[79]
Ru/BC-1	718	22.8	[80]
Ru/NPC	813	24.95	[81]
Ru(0)/SiO_2-$CoFe_2O_4$	172	45.6	[82]
Ru@SiO_2	200	38.2	[83]
Ru(1)@S1B-C10	202.4	24.13	[84]

Table 2. Cont.

Catalyst	TOF (mol$_{H2}$·mol$_{Ru}^{-1}$·min^{-1})	Ea (kJ mol^{-1})	Reference
Ru@SBA-15	316	34.8	[85]
Fe$_3$O$_4$@SiO$_2$-NH$_2$-Ru	617	15.05	[86]
Ru@Al$_2$O$_3$	83.3	-	[87]
Ru/γ-Al$_2$O$_3$	256	-	[88]
Ru/Al$_2$O$_3$-NFs	327	36.1	[89]
Ru(0)/TiO$_2$	241	70	[90]
Ru^0/HfO$_2$	170	65	[91]
Ru/MIL-96	231	47.7	[92]
Ru@MIL-53(Cr)	260.8	28.9	[93]
Ru@MIL-53(Al)	266.9	33.7	[93]
Ru/MIL-53(Al)-NH$_2$	287	30.5	[94]
Ru/PAF-72	294	-	[95]
Ru/Mg$_2$Al-LDH-h	85.7	50.3	[96]
Ru/Mg$_1$Al$_1$-LDHs	137.1	30.8	[97]
CF-BT-Ru	322	32.41	[71]

2. Hydrolytic Dehydrogenation of Ammonia Borane (AB) over Carbon Material-Supported Catalysts

Several catalytic supports have been explored for the synthesis of highly efficient catalysts for the hydrolytic dehydrogenation of AB. Carbon materials are one of the most intensively studied and the resulting catalysts have given very interesting results.

Akbayrak and Özkar explored the performance of catalysts supported on multi-walled carbon nanotubes (Ru(0)@MWCNT) by evaluating the activity of *in-situ* formed Ru nanoparticles (NPs) [72]. The resulting NPs had an average particle size range of 1.4–3.0 nm and were well-dispersed on the support. The effect of the Ru content was evaluated by checking the activity (in mL of H$_2$/min) of catalysts with metal contents of 0.73, 1.47, 1.91, 2.26, and 2.83 wt. %. Among those investigated, the sample with 1.91 wt. % displayed the best activity, with a TOF = 329 min^{-1} (mol$_{H2}$·mol$_{Ru}^{-1}$·min^{-1}). That catalyst was evaluated during four consecutive reaction cycles and it preserved 41% of its initial activity. Doe et al. also checked the performance of MWCNT-supported Ru catalysts with Ru(NH$_3$)$_6$Cl$_3$ as the metal precursor and using electrostatic adsorption (EA) and incipient wetness impregnation (IWI) methods [98]. Additionally, catalysts based on activated carbon and SiO$_2$ were prepared as reference materials. The Ru NPs were located on the external surface after both EA and IWI, and a smaller average size was achieved for the EA (2 and 3 nm for Ru/MWCNT-EA and Ru/MWCNT-IWI, respectively). Checking the performance of three sets of catalysts supported on MWCNTs, activated carbon, and SiO$_2$ with various average NPs size, it was observed that in all cases the catalysts with larger NPs attained higher reaction rates (expressed as turnover rates in mol$_{H2}$·mol$_{surface\ Ru}^{-1}$·s^{-1}), and higher TOF values were achieved for the Ru/MWCNT catalysts. Among those evaluated in that study, Ru/MWCNTs-EA produced the highest initial TOF value and the lowest activation energy, which was attributed to the hydrogen spillover taking place on metal NPs supported on CNTs.

Cheng et al. reported a simple method for the preparation of a Ru/graphene catalyst synthesized from graphene oxide and RuCl$_3$ using a one-step co-reducing approach with methylamine borane (MeAB) [68]. The resulting catalyst was compared to those synthesized by using different reducing agents, namely AB and NaBH$_4$. It was observed that the sample prepared with MeAB provided better results than those using AB or NaBH$_4$, which was attributed to a better control over nucleation and growth processes. It was determined that the average NPs size was 1.2, 1.7, and 2.0 nm for the catalysts reduced with MeAB, AB, and NaBH$_4$, respectively, suggesting that the size of the NPs increased as the reducing agent became stronger. The catalyst reduced with MeAB had a TOF of 100 mol$_{H2}$·mol$_{Ru}^{-1}$·min^{-1}, and activation energy of 11.7 kJ mol^{-1}, and it retained 72% of its initial activity after four reaction cycles. The same research group followed a very similar one-step co-reducing approach using ascorbic acid to synthesize Ru/graphene catalysts, which resulted in much higher TOF values of 600 $mol_{H2} \cdot mol_{Ru}^{-1} \cdot min^{-1}$ and activation energy of 12.7 kJ mol^{-1} [78], which pointed out the importance of the reducing agent in controlling the final catalytic performance. The high TOF value achieved by Ru/graphene was attributed to the narrow size distribution of the Ru NPs and the utilization of graphene as a suitable support.

In the study of Ma and co-workers, [75] ligand-free Ru NPs supported on carbon black were prepared *in-situ* from the reduction of the metal precursor (i.e., $RuCl_3$) by AB concomitantly with its hydrolysis. The resulting catalyst had an average nanoparticle size of 1.7 nm and showed a TOF of 429.5 $mol_{H2} \cdot mol_{Ru}^{-1} \cdot min^{-1}$. In that case, the reusability of the catalyst was checked during five consecutive reaction runs, after which it preserved 43.1% of its initial activity. The results of the characterization of the spent catalyst indicated that there were neither metal leaching nor aggregation of the NPs, so that the activity loss was attributed to an increasing concentration of the reaction products (i.e., metaborate and Cl^- ions) and their adsorption on the surface of the NPs. Furthermore, it was postulated that the increase in the viscosity in the solution after several reaction cycles could impede the diffusion of the reactant molecules and their collision with the Ru active sites.

Sun et al. studied the performance of poly(N-vinyl-2-pyrrolidone) (PVP)-stabilized Ru NPs loaded onto bamboo leaf-derived porous carbon (Ru/BC) [80]. In that case, the NPs were synthesized by *in-situ* reduction with AB using $RuCl_3 \cdot nH_2O$ as a metal precursor and with various PVP content (i.e., 0, 1, 3, 5, or 10 mg), which was used to avoid the agglomeration of the NPs. The catalysts with the best activity among those investigated (i.e., Ru/BC stabilized with 1 mg of PVP) displayed a TOF of 718 $mol_{H2} \cdot mol_{Ru}^{-1} \cdot min^{-1}$, and retained nearly 56% of the initial catalytic activity after 10 consecutive reaction cycles. That stability was much better than that of the PVP-free catalyst, which indicated the important role of PVP in enhancing the recyclability of the catalysts by preventing the agglomeration of the NPs. The same research group also investigated the performance of nitrogen-doped (N-doped) porous carbon materials [81]. The support was prepared by hydrothermal treatment of hydrochloride semicarbazide and glucose, and it was subsequently loaded with the metal precursor. The resulting Ru/NPC catalyst completed the dehydrogenation reaction in 90 s at room temperature, reaching a TOF of 813 $mol_{H2} \cdot mol_{Ru}^{-1} \cdot min^{-1}$. The stability of the catalyst was checked by performing five consecutive runs, after which 67.3% of the initial activity was preserved, and the activity decay was attributed to the agglomeration of the NPs.

Yamashita et al. addressed the preparation of highly efficient Ru/carbon catalysts prepared by pyrolysis of a supported Ru complex (i.e., tri(2,2-bipyridyl) ruthenium (II) chloride hexahydrate) [79]. In that study, the Ru catalysts were prepared by impregnating a commercial activated carbon with the metal precursor and subsequent decomposition of the metal complex by carrying out a heat treatment at temperatures ranging from 600 to 1000 °C (catalysts denoted as Ru/C(600)-(1000)). Transmission Electron Microscopy (TEM) analysis confirmed the formation of Ru NPs even for the lowest temperatures used for the decomposition of the metal complex. The average NP size strongly depended on the decomposition temperature and ranged from 3.8 to 13.5 nm for the Ru/C catalysts. Two additional reference samples were also synthesized: Ru/C(imp) prepared by the same protocol using $Ru(NO)(NO_3)_3$ as the metal precursor, and the Ru/SiO_2 catalyst prepared from $Ru(bpy)_3^{2+}$ and commercial fumed silica, with NP size of 3.3 and 2.0 nm, respectively. The results of the catalytic activity indicated that complete conversion of AB was attained with Ru/C(800), Ru/C(900) and Ru/C(1000), whereas it was not achieved with Ru/C(600), Ru/C(700), Ru/C(imp), and Ru/SiO_2. That observation led the authors to conclude that under the experimental conditions used, relatively large NPs were preferred for the reaction. Among those investigated, Ru/C(800) showed the most promising activity with an average TOF number of 670 $mol_{H2} \cdot mol_{Ru}^{-1} \cdot min^{-1}$. That catalyst achieved 100% of conversion even after four consecutive reaction runs, but the reaction rate was progressively sluggish. A summary of the results of the catalytic activity of the materials assessed in that work is plotted in Figure 3. It was also concluded that the electronic properties of Ru NPs played an important role in controlling the catalytic performance. It was claimed that Ru/C(800) had an optimum proportion of oxidized Ru species, which are important in the reaction.

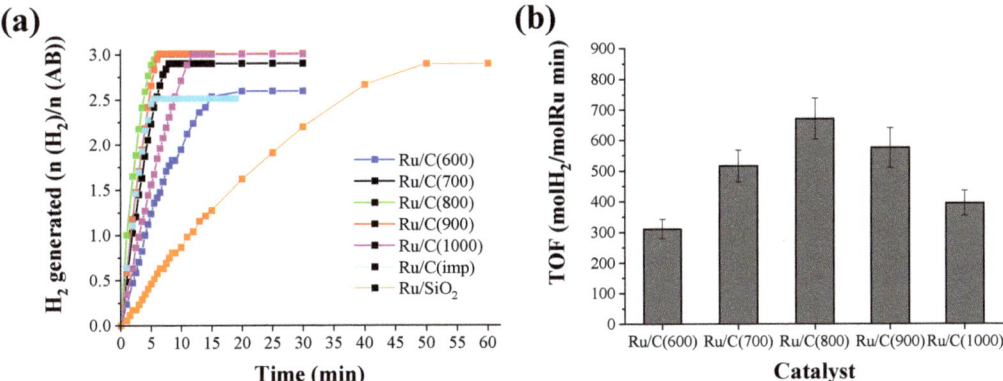

Figure 3. (a) Catalytic activity of the Ru-based catalysts in the AB hydrolysis reaction at 30 °C. (b) Turnover frequency (TOF) number ($mol_{H2} \cdot mol_{Ru}^{-1} \cdot min^{-1}$) calculated at t = 5.5 min for the Ru/C samples prepared by Ru complex decomposition. The average TOF numbers were calculated after performing three different catalytic tests. Adapted with permission from [79].

Gao et al. [70] studied the performance of Ru/nanodiamonds in the hydrolytic dehydrogenation of AB. That study was motivated by a wide variety of oxygen functional groups present in the nanodiamonds, which were expected to serve as anchoring points for the Ru NPs. Commercial nanodiamonds, with an average diameter of 5–10 nm were used as the support and $RuCl_3$ as the metal precursor, for the synthesis of catalysts with a metal content of 3.22, 4.82, 6.21, and 8.05 wt. %. The TOF values showed a volcano-type tendency, the best catalyst being that with a Ru content of 6.21 wt. % (229 $mol_{H2} \cdot mol_{Ru}^{-1} \cdot min^{-1}$). The recyclability tests performed with that sample indicated that total conversion of AB was achieved during four consecutive runs, but the TOF values decreased during the cycles, which was related to the increase of the NPs from 3.7 to 5.1 nm and the increasing viscosity and concentration of metaborate in the reaction solution.

Most of the catalysts used for the hydrolytic dehydrogenation of AB displayed good reusability and attained total conversion during several consecutive reaction runs. However, the reaction is frequently sluggish and longer reaction times are needed to produce 3 equivalents of H_2 per mole of AB, so the poor stability of the catalysts under reaction conditions is one of the most addressed issues.

The stability of the catalysts was shown to improve upon utilization of supports with abundant surface functional groups, which are known to increase the stability and reusability of the catalysts [8]. That was the case of Fan et al., who developed catalysts consisting of ultrafine and highly dispersed Ru NPs supported on N-doped carbon nanosheets formed by a hierarchically porous carbon material (HPCM) [76]. Ru/HPCM had an average NP size of 1.41 nm and narrow size distribution, which ranged from 0.6 nm to 2.0 nm. Such ultrafine NPs were confined into the micropores and mesopores of the support, affording numerous active sites and stabilizing the NPs from sintering under reaction conditions. X-ray Photoelectron Spectroscopy (XPS) analysis confirmed the presence of pyridinic, pyrrolic, and graphitic nitrogen. The importance of both pyridinic and graphitic nitrogen in enhancing catalytic performance was pointed out in that study. Ru/HPCM was evaluated during eight consecutive reaction cycles, after which it retained 50% of its initial activity. After that, the average NP size of the spent catalyst slightly increased to 1.47 nm and the partial activity decay was related to the catalyst loss and the passivation effect of metaborate ions formed along the recycling tests. The same group also synthesized Ru-based catalysts supported on N-doped bagasse-derived carbon materials (BC-hs) [73]. That biomass residue, which has abundant negative oxygen and nitrogen functional groups, was suitable for the interaction with Ru^{3+} cations of the metal precursor. Catalysts with various metal contents (i.e., 2.5, 3.5, 4.5, and 5.5 wt. %) were synthesized by *in-situ* reduction with AB, achieving homogeneously dispersed ultrafine Ru NPs. It was observed that the best-performing catalyst preserved 80% of its TOF value after five runs, demonstrating the suitability of the BC-hs support to stabilize the metal NPs. The partial loss of the activity was attributed to changes in the NP size and loss of the catalyst during the separation and washing steps.

Ma et al. used a support based on a N-doped porous carbon material (NC-Fe) using a facile pyrolysis of a porous organic polymer (POP) synthesized from ferrocene carboxaldehyde and melamine as the starting materials [69]. The resulting Ru catalyst (Ru/NC-Fe) achieved the total conversion of AB during five cycles, but the reaction was considerably sluggish. That catalyst contained γ-Fe_2O_3, so it was easily recovered with a magnet. The activity decay was attributed to both the metaborate formed in the reaction and the agglomeration of the metal NPs.

Liu et al. used a N-enriched hierarchically macroporous-mesoporous carbon support that was synthesized by a co-template evaporation-induced self-assembly approach with SiO_2 nanospheres as a macroporous hard template and F127 as a mesoporous soft template and a soft nitriding by the low-temperature thermolysis of urea [99]. The resulting support, denoted as hPCN, was loaded with Ru NPs (Ru@hPCN), and reference catalysts based on pure macroporous and pure mesoporous support were also prepared (Ru@macroPCN and Ru@mesoPCN, respectively). Ru@hPCN exhibited the best activity and total dehydrogenation was finished in 120 s, while longer reaction times were required for the reference samples and for a commercial Ru/C catalyst (160, 280, and 880 s, for Ru@macroPCN, Ru@mesoPCN, and Ru/C, respectively). Such better performance of Ru@hPCN was also evidenced by the higher TOF values achieved at 60 °C (1850, 1258, 902, and 308 min^{-1}, for Ru@hPCN, Ru@macroPCN, Ru@mesoPCN, and Ru/C, respectively). The reusability of Ru@hPCN was evaluated during four consecutive reaction cycles and it was observed that the NP size increased from 0.7 nm to 2.7 nm. Additionally, the amount of surface N species decreased from 13.4 at. % to 2.3 at. %, which suggested that the N-enriched species contributed to improving the catalytic performance by dissociating the electropositive $H^{\delta+}$ from water molecules and the breakage of the B-N bonds.

Not only N-doped carbon materials, but also other N-containing carbon-based supports were shown to be effective in stabilizing Ru NPs for the hydrolytic dehydrogenation of AB. For instance, Zheng et al. used hierarchical porous graphitic carbon nitride (g-C_3N_4) nanosheets to anchor ultrafine Ru NPs [100]. In that case, the supports were prepared from melamine and various amounts of NH_4Cl as a dynamic gas template. The NPs were encapsulated into the network of the g-C_3N_4 by the reduction of the metal precursor with $NaBH_4$ to achieve a final metal content of 1.91 wt. %. The characterization of the catalyst indicated that the resulting NPs had a size that ranged from 1.9 to 5.1 nm and they exhibited uniform dispersion onto the support. Among the synthesized supports, that prepared with a melamine to NH_4Cl mass ratio of 1:3 exhibited the highest surface area (S_{BET} of 59 $m^2\,g^{-1}$) and its counterpart Ru catalyst was selected to assess the performance in the dehydrogenation of AB. That catalyst displayed satisfactory recyclability even after four consecutive reaction cycles. The activity loss observed from the fifth cycle was attributed to a particle aggregation of the NPs and to the accumulation of NH_4BO_2 species in the reaction solution, which increases its viscosity and blocks the active sites of the catalyst.

Tang et al. also explored the suitability of Ru/g-C_3N_4 catalysts for the dehydrogenation of AB [77]. In that case, the catalysts were prepared from urea and $RuCl_3$. Catalysts with various metal loadings (i.e., 4.10, 3.28, 2.46, and 1.64 wt. %) were synthesized. The time needed for the reaction to be completed decreased to 3.5 min for the catalyst with 3.28 wt. %, which showed the highest TOF value (459.3 $mol_{H2}\cdot mol_{Ru}^{-1}\cdot min^{-1}$). Concerning the recyclability of Ru/g-C_3N_4, it preserved 50% of the initial catalytic activity after the fourth run, and the activity loss was related to the increase of the size of the NPs from 2.8 nm to 4.1 nm and the adsorption of B species on the surface of the NPs.

Yamashita et al. also studied the performance of g-C_3N_4 supported Ru catalysts [101]. In that case, carbon/g-C_3N_4 composites with various carbon contents (C(x)/g-C_3N_4; "x" is the initial carbon weight per cent of 0.1, 0.5, 1.0, 2.0, and 4.0 wt. %, respectively) were synthesized from glucose and dicyandiamide by a simple experimental procedure. The obtained supports were subsequently impregnated with $RuCl_3 \cdot 3H_2O$ to obtain Ru NPs after reduction with H_2 gas at 300 °C. The incorporation of carbon was reported to extend the absorption of the materials to the visible region of 480–800 nm compared to the pristine g-C_3N_4, so the resulting Ru/C(x)/g-C_3N_4 were interesting photocatalysts for the dehydrogenation of AB under visible-light irradiation. Additionally, the incorporation of carbon served to achieve smaller Ru NPs than those achieved in the raw g-C_3N_4, which also affected the catalytic activity. Figure 4 contains information on the catalytic activity displayed by Ru/C(x)/g-C_3N_4. It was observed that the addition of moderated carbon contents in the catalysts enhanced the activity compared to that displayed by Ru/g-C_3N_4, which was related to the smaller NPs shown in those samples (Figure 4a). It was also seen that the reaction rate of all the materials improved under visible light irradiation (Figure 4b), achieving the fastest reaction rate with Ru/C(1.0)/g-C_3N_4.

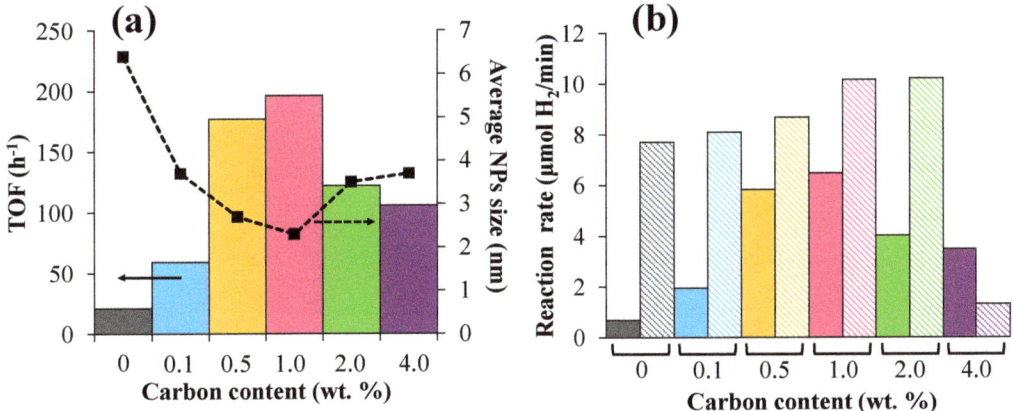

Figure 4. (a) TOF (h^{-1}) for Ru/C(x)/g-C_3N_4 as a function of the carbon content and average Ru NP size. (b) Initial reaction rate for the dehydrogenation of AB as a function of the carbon content under dark conditions (solid bars) and visible light irradiation (hollow bars). Adapted from [101].

Interesting results were also achieved by Fan et al., who synthesized phosphorus-doped carbon-supported Ru catalysts (Ru/PPC) [74]. In that study, the supports were prepared from hypercrosslinked polymer networks of triphenylphosphine and benzene, and they were subsequently used for the synthesis of Ru catalysts with various metal contents (i.e., 1.5, 2.5, 3.5, and 4.5 wt. %). Small and well-distributed Ru NPs were attained in all the materials, the smallest for Ru/PPC with 3.5 wt. % of Ru (average size of 1.13 nm). The catalytic activity was shown to be dependent on the metal content, achieving TOF values of 174, 325, 413, and 290 $mol_{H_2} \cdot mol_{Ru}^{-1} \cdot min^{-1}$ for catalysts with 1.5, 2.5, 3.5, and 4.5 wt. % of Ru, respectively. The stability of the most active material was evaluated during four consecutive cycles, observing that its activity gradually decreased after the first run. Such loss of activity was attributed to the NP sintering (from 1.13 to 2.47 nm) as well as the catalyst loss in the separation and washing steps.

3. Hydrolytic Dehydrogenation of AB over Oxide-Supported Catalysts

As in most catalytic reactions, carbon materials are the most fruitfully studied supports. However, interesting results have also been achieved with oxide-supported catalysts, with silica (SiO_2) and alumina (Al_2O_3) the most investigated. Representative examples of such systems are summarized in this section.

SiO_2 has been shown to serve as a suitable support for the dehydrogenation of AB. SiO_2, with various structures and morphologies, have been nicely utilized for the development of well-performing catalysts. For instance, Zhu et al. reported on Ru NPs confined in SBA-15 (Ru@SBA-15) by using a double solvent approach (with hexane and water) [85]. Metal loadings of 0.5, 1.1, 2.1, 3.2, and 4.0 wt. % were used, which resulted in NPs with an average size of 2.0 ± 0.6 nm, 2.2 ± 0.6 nm, 3.0 ± 0.8 nm, and 3.7 ± 0.7 nm, respectively. Among investigated, Ru@SBA-15 with a metal content of 2.1 wt. % displayed the best activity, which was also superior to those catalysts with non-confined NPs supported on SBA-15 and SiO_2. It also showed good durability after five consecutive reaction runs at room temperature.

Chen et al. designed Ru catalysts supported on cubic 3D cage-type mesoporous silica SBA-1 functionalized with carboxylic acid (Ru/S1B-C10) [84]. The carboxylic acid was incorporated by co-condensation of tetraethyl orthosilicate (TEOS) and carboxyethylsilanetriol sodium salt in the presence of poly(acrylic acid) and hexadecylpyridinium chloride. The –COOH groups were uniformly distributed within the mesopores of the support, which assisted the preparation of well-dispersed and small Ru NPs. Catalysts with metal loadings from 0.5 to 2.0 wt. % were synthesized with both –COOH free support (SIB-C0) and –COOH containing support (S1B-C10). It was observed that while the average NP size observed in Ru/S1B-C0 catalysts increased from 2.8 to 4.6 nm when the metal loading increased from 0.5 to 2.0 wt. %, it only increased from 2.0 to 3.2 nm in Ru/SIB-C10 for similar metal contents. Catalysts with 1 wt. % of Ru displayed better performance for both SIB-C0 and SIB-C10 supports than materials with other Ru loadings. Among those investigated, Ru(1)@S1B-C10 had the highest TOF value (202 $mol_{H2} \cdot mol_{Ru}^{-1} \cdot min^{-1}$), which was attributed to the nanosized Ru particles and their good dispersion as well as the effect of the pore confinement of the support. The reusability of the best-performing catalysts was evaluated during five reaction cycles and even though the total conversion was achieved in all cases, the reaction rates decreased with an increase in the number of cycling tests, which was related to a partial metal leaching and restricted access of reactants to the Ru active sites originated by the adsorption of metaborate on the surface of the NPs.

Yao el at. prepared core-shell Ru@SiO_2 catalysts with various metal contents (i.e., 1.0, 2.0, 3.0, 4.0, 6.0, 8.0, and 10.0 wt. %), which consisted of Ru NPs of ~2 nm embedded in the center of spherical SiO_2 particles of ~25 nm [83]. Among those investigated, Ru@SiO_2 with 6 wt. % of Ru loading exhibited the best performance for the production of H_2 from AB, with a TOF of 200 $mol_{H2} \cdot mol_{Ru}^{-1} \cdot min^{-1}$. A much slower reaction rate was observed for a supported Ru/SiO_2 catalyst used as a reference sample, which was related to the easy aggregation occurring in Ru/SiO_2. The Ru@SiO_2 core-shell catalyst displayed good recycling stability for five cycles, but the reaction was progressively sluggish during the cycles.

Özkar et al. explored the activity of catalysts formed by Ru NPs loaded on magnetic silica-coated cobalt ferrite (Ru(0)/SiO_2-$CoFe_2O_4$) [82]. That catalyst showed a moderated activity, with a TOF of 172 $mol_{H2} \cdot mol_{Ru}^{-1} \cdot min^{-1}$ at room temperature, but excellent recyclability during 10 reaction runs, preserving 94% of the initial activity. After each reaction cycle, the spent catalyst was isolated using a permanent magnet, and no metal leaching was detected.

Onat et al. [86] also developed SiO_2-based magnetic core-shell catalysts. In that study, Ru NPs were loaded on amino functionalized silica-covered magnetic NPs (Fe_3O_4@SiO_2-NH_2-Ru), and the resulting catalysts were evaluated in the dehydrogenation of AB. It was claimed that the amine group served to increase the electron transfer from the core-shell Fe_3O_4@SiO_2 structure to the surface of the Ru NPs, which resulted in enhanced catalytic activity in the production of H_2. The developed catalyst had a TOF value of 617 $mol_{H2} \cdot mol_{cat}^{-1} \cdot min^{-1}$ and showed good stability during eight reaction cycles.

Al$_2$O$_3$ has also been applied for the synthesis of Ru-based catalysts for the dehydrogenation of AB. For instance, Metin el al. reported an easy method for the synthesis of nearly monodisperse Ru NPs, which consisted in the thermal decomposition and simultaneous reduction of the metal precursor (ruthenium(III) acetylacetonate (Ru(acac)$_3$) in the presence of oleylamine (serving as a stabilizer and reducing agent) and benzylether (used as solvent) [87]. The as-synthesized NPs (with an average size of 2.5 nm) were subsequently loaded on γ-Al$_2$O$_3$ to build a Ru@Al$_2$O$_3$ catalyst with 1 wt. % of Ru. That material had a TOF of 39.6 mol$_{H2}$·mol$_{Ru}^{-1}$·min^{-1}. It was observed that the TOF value increased to 83.3 mol mol$_{H2}$·mol$_{Ru}^{-1}$ after carrying out a treatment of the catalyst with acetic acid, which removed the organics from the surface of the NPs. The treated catalysts exhibited great stability even after 10 consecutive reaction runs, and the TOF only decreased 10% of the initial value.

Chen el at. developed a microporous crystalline γ-Al$_2$O$_3$ with a large surface area (large surface, 349 m^2 g^{-1}) prepared from a microporous covalent triazine framework (CTF-1, surface area of 697 m^2 g^{-1}) as a template [88]. Ru-based catalysts with 1, 2, and 5 wt. % were prepared from RuCl$_3$ and they were evaluated in the dehydrogenation of AB. The high surface area and the hierarchical pore structures of the micropores developed in the synthesized γ-Al$_2$O$_3$ was claimed to be responsible for the better performance achieved by the materials studied in that work compared to those reported elsewhere for Ru/γ-Al$_2$O$_3$ catalysts (TOF as high as 256.8 mol$_{H2}$·mol$_{Ru}^{-1}$·min^{-1} was obtained for Ru/γ-Al$_2$O$_3$ with 2 wt. %; while values of one order of magnitude lower were attained for Ru/γ-Al$_2$O$_3$ reported in other studies).

Fan et al. evaluated the performance of catalysts formed by Ru NPs supported on Al$_2$O$_3$ nanofibers with an average length of 200–300 nm and a BET surface area of 300 m^2 g^{-1} (Ru/Al$_2$O$_3$-NFs) [89]. Catalysts with metal contents from 2.52 to 4.91 wt. % were prepared in-situ by reducing the metal precursor with AB. It was observed that the time needed to achieve a total conversion of AB decreased with increasing the metal content. The stability test indicated that AB conversion was completed even during five consecutive runs, but the reaction rate decreased along the cycles, which was attributed to an increase of the average NPs size from 2.9 to 3.1 nm.

Lee et al. investigated the effect of the crystal phase of Ru on the dehydrogenation of AB by using face-centered cubic (fcc) structures and hexagonal close-packed (hcp) structured Ru NPs loaded on γ-Al$_2$O$_3$ [102]. Catalysts with 1 wt. % of Ru and different sizes of both fcc and hcp Ru NPs were prepared. The size and the crystal phase were controlled by adjusting the experimental conditions, in terms of amount and type of metal precursor, solvent, and amount of PVP. It was observed that under the experimental conditions used in that study, fcc NPs were achieved using Ru(acac)$_3$ while RuCl$_3$ originated hcp NPs. XPS analysis suggested that fcc Ru NPs were more easily oxidized than the hcp counterpart since a larger relative proportion of Ru^{4+} was detected (i.e., [Ru^{4+}]/[Ru^0] of 28.2 and 19.2% for fcc and hcp Ru/γ-Al$_2$O$_3$, respectively). The catalytic activity of fcc and hcp NPs with different sizes (i.e., 2.4, 3.5, 3.9, and 5.4 nm) was evaluated by monitoring the H$_2$ generation profiles at 25 °C (see Figure 5). As can be seen, hcp NPs displayed better performance regardless of the size of the NPs. It was also seen that the difference between hcp and fcc NPs became smaller as the NPs increased. Additionally, the opposite tendency was seen for fcc and hcp NPs: the catalytic activity of fcc Ru/γ-Al$_2$O$_3$ enhanced with increasing NP size, while hcp displayed worse performance with increasing sizes. Density functional theory (DFT) calculations were done to determine the adsorption energy of O$_2$ molecules on the (001) crystal plane of fcc and hcp Ru to get information about their easy oxidation. The results obtained suggest that the fcc Ru was easily oxidized than hcp Ru, which was consistent with the experimental results observed in that study. Thus, the authors of that study ascribed the worse performance of smaller NPs with fcc crystal phases to their higher degree of oxidation, while the tendency observed for hcp was attributed to size effects.

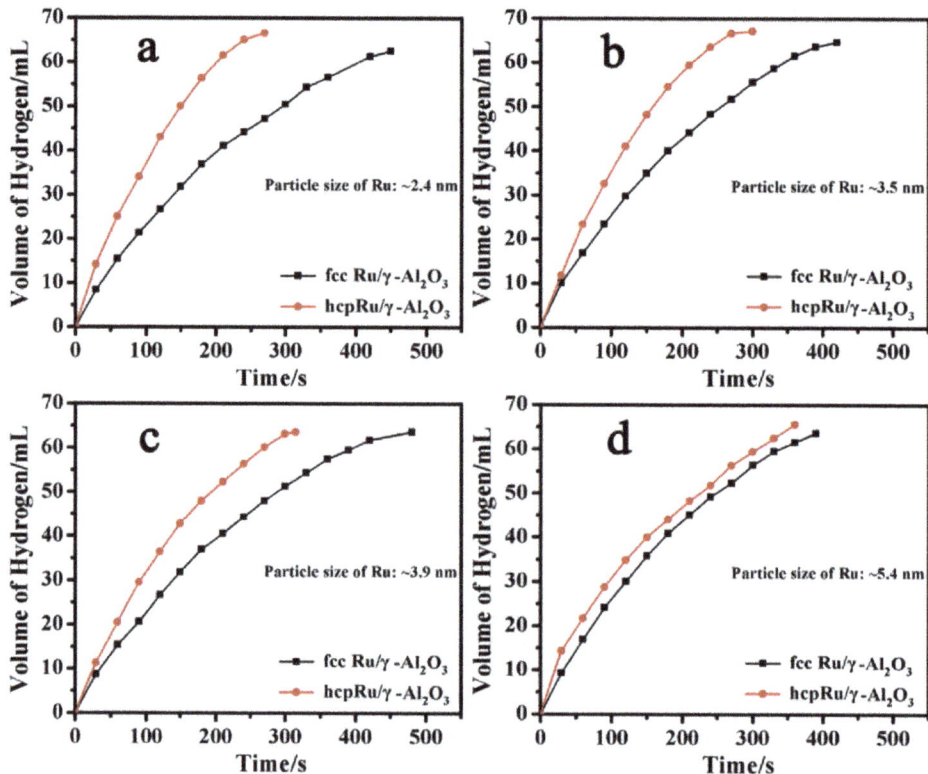

Figure 5. The plot of H$_2$ volume (mL) versus time (s) graph for the dehydrogenation of AB (n$_{AB}$ = 1 mmol) catalyzed by the Ru with different crystal phase (fcc: face-centered cubic; hcp: hexagonal close-packed) for different particle sizes: (**a**) ~2.4 nm, (**b**) ~3.5 nm, (**c**) ~3.9 nm, (**d**) ~5.4 nm. Experimental conditions: 25 °C and n$_{Ru}$/n$_{AB}$ = 0.003. Adapted from [102].

Some examples of the use of other oxides such as titania (TiO$_2$) [90,103], ceria (CeO$_2$) [104,105], and hafnia (HfO$_2$) [91] as catalytic supports of Ru NPs for the hydrolytic dehydrogenation of AB can also be found in the literature.

4. Other Supported Ru Catalysts

Carbon materials and oxides are the most investigated supports for the hydrolytic dehydrogenation of AB, but there are other interesting materials that have also attracted great interest in the last years. Among them, the utilization of MOF should be highlighted because of the attention drawn not only for this application, but for many other catalytic reactions [106–108]. They have been shown to be suitable to embed metal NPs as well as to stabilize them by incorporating additional functionalization. This was the case of Chen et al., who achieved well-dispersed Ru NPs immobilized within the pores of amine-functionalized MIL-53 by using an *in-situ* impregnation-reduction method [94]. It was claimed that the amino groups were located at the acid linkers in MIL-53(Al)-NH$_2$, and served as Lewis bases, thus stabilizing the Ru precursor (i.e., RuCl$_3$) during the impregnation step. An amino-free catalyst was prepared for comparison purposes. The average NP size was determined to be 1.22 nm, smaller than the mean diameter of the pores of the amino functionalized MIL-53(Al) (2.07 nm) so that they were embedded in the framework of the MOF, while larger NPs were located on the external surface. Ru/MIL-53(Al)-NH$_2$ catalyst displayed better activity than the amino-free counterpart, and also better durability and reusability. Such better performance exhibited by Ru/MIL-53(Al)-NH$_2$ was attributed to the amino groups, which assisted the formation and stabilization of small Ru NPs. Xia et al. also used MIL-53 as support of Ru NPs for the hydrolytic dehydrogenation of AB. In that study, Ru NPs were deposited on MIL-53(Cr) and MIL-53(Al) by the impregnation method with $RuCl_3$. Catalysts with Ru contents of 0.19, 0.67, 1.61, and 2.65 wt. % were obtained for Ru@MIL-53(Cr), and metal contents of 0.12, 0.74, 1.95, 2.59 wt. % were achieved for Ru@MIL-53(Al). Among the samples assessed, 2.65 wt. % Ru@MIL-53(Cr) and 2.59 wt% Ru@MIL-53(Al) displayed the best performance, with TOF values of 260.8 and 266.9 $mol_{H2} \cdot mol_{Ru}^{-1} \cdot min^{-1}$, respectively. Those catalysts also showed good stability, preserving 71% and 75% of the initial catalytic activity of Ru@MIL-53(Cr) and 2.59 wt. % Ru@ MIL-53(Al) after the fifth run, respectively. Chen et al. prepared catalysts formed by ultrasmall Ru NPs supported on MIL-96 (Ru/MIL-96) [92]. The MOF selected in that study was claimed to be a 3D framework with three different kind of cages as well as with thermal and chemical stability in water. Ru catalysts loaded on other supports (i.e., carbon black, SiO_2, γ-Al_2O_3, and GO) were prepared for comparison purposes. Ru/MIL-96 was the most active catalysts checked in that study, with a TOF of 231 $mol_{H2} \cdot mol_{Ru}^{-1} \cdot min^{-1}$. However, that material did not show suitable stability, since it only retained 65% of its initial activity after the fifth run, which was attributed to the increasing NP size and viscosity of the solution.

Zhu et al. explored the suitability of a N-containing microporous organic framework (POF) as a scaffold to anchor Ru NPs [95]. The resulting catalyst, which was denoted as Ru/PAF-72, has an average NP size of 1–2 nm and showed a TOF of 294 $mol_{H2} \cdot mol_{Ru}^{-1} \cdot min^{-1}$. The most remarkable aspect of that system was its great stability even after 10 consecutive reaction cycles. For comparison purposes, the stability of a catalyst based on Ru NPs loaded onto carbon black was also evaluated, showing a significant activity decay after only four cycles.

Additionally, some other supports, which are less frequently used in catalysts, have served to develop Ru catalysts for the hydrolytic dehydrogenation of AB. For instance, Cai et al. developed a sophisticated catalyst based on Ru NPs loaded a natural polyphenolic polymer (bayberry tannin, BT) immobilized on collagen fibers (CF). The catalysts, denoted as CF-BT-Ru, were synthesized by immobilization of BT, crosslinking of glutaraldehyde, and subsequent chelation of Ru^{3+} (final metal loadings of 0.58, 1.03, 1.57, and 2.12 wt. %) [71]. The procedure used is shown in Figure 6. Various reference samples were also prepared for comparison purposes (i.e., CF-BT-Ru, CF-Ru, and Ru-carbon material (Ac, GO, CNTs and g-C_3N_4)). It was observed that the presence of BT enhanced the dispersion of the metal active phase, due to the interaction with the phenolic groups, so that the BT containing catalyst has much smaller Ru NPs than the BT-free counterpart (i.e., 2.6 ± 0.6 and 6.5 ± 0.5 nm, respectively). XPS characterization indicated that the catalyst contained positively charged nitrogen and neutral amine groups. Such neutral amine groups were responsible for the stabilization of the Ru NPs by providing some of their electrons. The six catalysts evaluated showed total conversion of AB, but different reaction rates, with CF-BT-Ru being the most active with a TOF of 322 $mol_{H2} \cdot mol_{Ru}^{-1} \cdot min^{-1}$.

In an attempt to gain insights into the less explored effect of the composition of the support, Zhao et al. [96] selected a material that consisted of composition-adjustable layered double hydroxide (MgAl-LDHs) as support for Ru NPs. The general formula of the selected support is $[M^{2+}_{1-x}M^{3+}_x(OH)_2]^{x+}[A^{n-}]_{x/n} \cdot yH_2O$, where M^{2+} and M^{3+} are cations, and A^{n-} is the charge balancing anion. In that study, supports with different compositions of MgAl-LDHs were synthesized by urea hydrolysis (supports denoted as Mg_2Al-LDH-h, Mg_3Al-LDH-h, and Mg_4Al-LDH-h). Mg_4Al-LDH was also prepared from a co-precipitation method (support denoted as Mg_4Al-LDH-p). After that, Ru catalysts were prepared from $RuCl_3 \cdot H_2O$. The results of the catalytic activity indicated that the activity followed the order Ru/Mg_2Al-LDH-h > Ru/Mg_3Al-LDH-h > Ru/Mg_4Al-LDH-(h) > Ru/Mg_4Al-LDH-(p), with TOF values of 85.7, 63.3, 42.7, and 40.1 $mol_{H2} \cdot mol_{Ru}^{-1} \cdot min^{-1}$, respectively. According to that observation, it was postulated that the Mg/Al ratio, which is related to the relative acidity of the material and the support-Ru interaction, was a key aspect in controlling the final catalytic performance. It was found that Ru/Mg_2Al-LDH-h had more Brønsted acid sites and it showed a weaker interaction with Ru species, which enabled the existence of Ru in the metallic state, thus explaining the better activity exhibited by that catalyst. That sample preserved its stability during the first four cycles, but it slightly decayed from the fourth to the tenth consecutive run, which was attributed to the increased viscosity of the solution, which impedes the diffusion of AB molecules, thus hindering their interaction with the Ru active sites.

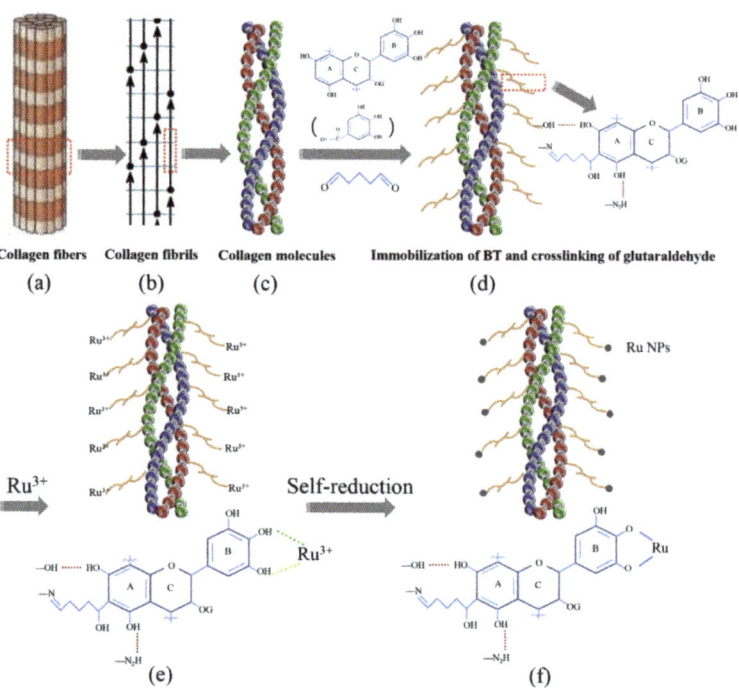

Figure 6. Diagram of (**a**) collagen fibers, (**b**) collagen fibrils, (**c**) collagen molecules, (**d**) immobilization of bayberry tannin (BT), and crosslinking of glutaraldehyde, (**e**) chelation of Ru^{3+} ions with phenolic hydroxyl groups, and (**f**) formation of Ru NPs under self-reduction. Reprinted with permission from [71].

Sun et al. also explored the performance of MgAl-LDH-supported Ru catalysts for the hydrolytic dehydrogenation of AB [97]. In this case, supports with a composition of $Mg_{0.5}Al_1$-LDHs, Mg_1Al_1-LDHs, Mg_2Al_1-LDHs, and Mg_3Al_1-LDHs were prepared and loaded with Ru NPs. Among those samples, Ru/Mg_1Al_1-LDHs showed better activity, completing the hydrogen release in 130 s, compared to the other catalysts that needed longer reaction times (180, 140, and 230 s, for Ru/$Mg_{0.5}Al_1$-LDHs, Ru/Mg_2Al_1-LDHs, and Ru/Mg_3Al_1-LDHs). In this case, such tendency was also attributed not only to the higher content of Brønsted acid sites in Ru/Mg_1Al_1-LDHs, but also to the higher purity of Mg_1Al_1-LDHs, which did not contain boehmite (AlO(OH)) and hydromagnesite ($Mg_5(-CO_3)_4(OH)_2·4H_2O$). Ru/$Mg_1Al_1$-LDHs exhibited good stability during 10 reaction runs, and 58.1% of the initial activity remained after those cycles. DTF calculations were conducted to get information on the promotion effect of MgAl-LDHs catalysts. It was determined that Ru/MgAl-LDHs catalysts have the beneficial electronic properties to accelerate the H_2O dissociation to form O–H bonds, activating the H_2O molecules during the hydrolytic dehydrogenation of AB.

5. Conclusions and Perspectives

There is great hope for the potential of hydrogen as an energy vector, which motivates the search for alternatives that overcome the limitations that are frequently related to its storage. Chemical hydrogen storage stands up as a promising option and there are several hydrogen carrier molecules that afford satisfactory hydrogen capacity. Among them, ammonia borane has drawn much attention, and research to exploit the potential of ammonia borane as a hydrogen storage material has been intensified in the last years.

This review summarized some of the recent Ru-based heterogeneous catalysts applied in the hydrolytic dehydrogenation of ammonia borane. A perusal of the most frequently used catalysts is included, paying particular attention to those heterogeneous catalysts with carbon materials and oxides as supports. Among the vast diversity of supported Ru catalysts studied thus far, carbon material-based catalysts frequently attain the best performance.

Most of the investigations deal with the optimization of the properties of the active phase such as the nanoparticle size and morphology, while the effect of the composition of the support is less explored. Most of the catalysts experienced deactivation during few consecutive reaction cycles, which is linked to several factors: (i) aggregation of the nanoparticles; (ii) metal leaching; and (iii) accumulation of metaborate in the solution and change in the viscosity. It is expected that such phenomena involved in the deactivation of the catalysts could be partially averted by using encapsulated or nanoconfined metal catalysts upon selection of a suitable porous material serving as a host for the metal active phase. Most studies include information about the NP size and metal content of the spent catalysts, but no information has been reported about the concentration of metaborate species and viscosity of the solution before and after the reaction, which would help to verify such possible reasons for deactivation that are frequently mentioned but never confirmed. Some other interesting works evidenced the formation of different B-containing products such as $B(OH)_3$, BO_2^-, $B(OH)_4^-$, and polyborates [64,65,109–111].

Some studies have aimed at enhancing the cyclability of the catalysts by stabilizing the metallic phases using strong metal-support interaction, using, for example, a support with abundant functional groups. Nitrogen-containing supports have been widely studied, but the resulting catalysts are still lacking in stability during the cycles. Hence, the insufficient stability of the assessed catalysts is frequently the weakest point indicated in the literature for the heterogeneous catalysts used in the dehydrogenation of ammonia borane. The regeneration of the catalysts after reaction would therefore be a very interesting issue to be considered in future works.

No attention is paid to the actual cornerstone for the successful application of this hydrogen carrier molecule in the hydrogen storage scenario, which is the regeneration of ammonia borane from the products obtained in its decomposition reaction. The AB regeneration problem, which was pointed out long back [112], remains unsolved nowadays and studies dealing with the regeneration of AB are still sparse [113]. The nature of the by-products of the hydrolysis of AB, which are mainly borate species, complicates the regeneration of AB, since the stable B-O bond formed in the by-product are strong and they are not easily reconverted to the B-H bond present in AB molecules, so a strong reducing agent would be needed. Thus, multi-step reactions are required for the regeneration of AB. For instance, Liu et al. proposed a regeneration process that implied the conversion of boric acid to trimethyl borate ($B(OCH_3)_3$) by esterification with methanol. Then, $B(OCH_3)_3$ can react with NaH, generating $NaBH_4$ and, finally, AB was formed upon reaction of NaH with ammonia sulfate in THF [114]. Gagare et al. aimed at regenerating AB from $NH_4B(OMe)_4$ using $LiAlH_4$ as a reducing agent and NH_4Cl as an ammonia source [115]. Vasiliu et al. reported on the regeneration of AB from a more simplified and energy-efficient process that involved minimum reaction steps [116]. However, their starting point was not the regeneration of the by-products originating from the production of 3 equivalents of H_2 per molecule of AB, but polyborazylene was used, which is formed upon generation of two equivalents of H_2 per molecule of AB (partially spent AB). In that case, polyborazylene was converted to AB nearly quantitatively by 24 h treatment with N_2H_4 in liquid NH_3 at 40 °C. More recently, Sharma et al. [117] studied the regeneration of AB using a digestion-based approach in the presence of methanol and subsequent regeneration with the reducing agent (i.e., $LiAlH_4$).

There is still plenty of room for improvement in both the design and development of stable and reusable catalysts as well as in the processes involved in the real application of ammonia borane as a hydrogen storage material, especially for those on-board systems. One of the main issues to be tackled is the high cost of ammonia borane compared to other hydrogen storage systems, so finding cost-effective ways for the synthesis and regeneration of ammonia borane is highly desirable.

The present review contains only information on monometallic Ru-based catalysts, which have been shown to be the most effective to attain suitable catalytic behavior, but there are other compositions that have also displayed interesting results.

Author Contributions: M.N.-G. designed the structure of the review and wrote the manuscript; D.S.-T. contributed to the writing of the manuscript; and D.C.-A. reviewed the paper. All authors have read and agreed to the published version of the manuscript.

Funding: This work was financed by the MICINN, FEDER (RTI2018-095291-B-I00). MNG thanks the Plan GenT project (CDEIGENT/2018/027) for their financial support. DST thanks MICINN for the "Juan de la Cierva" contract (IJCI-2016-27636) and the Vicerrectorado de Investigación y Transferencia de Conocimiento de la Universidad de Alicante (GRE19-16).

Institutional Review Board Statement: Not applicable.

Informed Consent Statement: Not applicable.

Data Availability Statement: Data sharing not applicable.

Conflicts of Interest: The authors declare no conflict of interest.

References

1. Zheng, J.; Zhou, H.; Wang, C.-G.; Ye, E.; Xu, J.W.; Loh, X.J.; Li, Z. Current research progress and perspectives on liquid hydrogen rich molecules in sustainable hydrogen storage. *Energy Storage Mater.* **2021**, *35*, 695–722. [CrossRef]
2. McGee, M. Available online: https://www.co2.earth/ (accessed on 12 March 2021).
3. Mazloomi, K.; Gomes, C. Hydrogen as an energy carrier: Prospects and challenges. *Renew. Sustain. Energy Rev.* **2012**, *16*, 3024–3033. [CrossRef]
4. Abdin, Z.; Zafaranloo, A.; Rafiee, A.; Mérida, W.; Lipiński, W.; Khalilpour, K.R. Hydrogen as an energy vector. *Renew. Sustain. Energy Rev.* **2020**, *120*, 109620. [CrossRef]
5. Navlani-García, M.; Mori, K.; Kuwahara, Y.; Yamashita, H. Recent strategies targeting efficient hydrogen production from chemical hydrogen storage materials over carbon-supported catalysts. *NPG Asia Mater.* **2018**, *10*, 1–16. [CrossRef]
6. Lai, Q.; Sun, Y.; Wang, T.; Modi, P.; Cazorla, C.; Demirci, U.B.; Ares Fernandez, J.R.; Leardini, F.; Aguey-Zinsou, K.-F. How to Design Hydrogen Storage Materials? Fundamentals, Synthesis, and Storage Tanks. *Adv. Sustain. Syst.* **2019**, *3*, 1900043. [CrossRef]
7. Zhang, F.; Zhao, P.; Niu, M.; Maddy, J. The survey of key technologies in hydrogen energy storage. *Int. J. Hydrogen Energy* **2016**, *41*, 14535–14552. [CrossRef]
8. Salinas-Torres, D.; Navlani-García, M.; Mori, K.; Kuwahara, Y.; Yamashita, H. Nitrogen-doped carbon materials as a promising platform toward the efficient catalysis for hydrogen generation. *Appl. Catal. A Gen.* **2019**, *571*, 25–41. [CrossRef]
9. U.S. Department of Energy. Available online: https://www.energy.gov/eere/fuelcells/doe-technical-targets-onboard-hydrogen-storage-light-duty-vehicles (accessed on 12 March 2021).
10. Berenguer-Murcia, Á.; Marco-Lozar, J.P.; Cazorla-Amorós, D. Hydrogen Storage in Porous Materials: Status, Milestones, and Challenges. *Chem. Rec.* **2018**, *18*, 900–912. [CrossRef]
11. Schlapbach, L.; Züttel, A. Hydrogen-storage materials for mobile applications. *Nature* **2001**, *414*, 353–358. [CrossRef] [PubMed]
12. Demirci, U.B.; Miele, P. Chemical hydrogen storage: "Material" gravimetric capacity versus "system" gravimetric capacity. *Energy Environ. Sci.* **2011**, *4*, 3334–3341. [CrossRef]
13. Rivard, E.; Trudeau, M.; Zaghib, K. Hydrogen storage for mobility: A review. *Materials* **2019**, *12*, 1973. [CrossRef]
14. Dalebrook, A.F.; Gan, W.; Grasemann, M.; Moret, S.; Laurenczy, G. Hydrogen storage: Beyond conventional methods. *Chem. Commun.* **2013**, *49*, 8735–8751. [CrossRef] [PubMed]
15. Yüksel Alpaydın, C.; Gülbay, S.K.; Ozgur Colpan, C. A review on the catalysts used for hydrogen production from ammonia borane. *Int. J. Hydrogen Energy* **2020**, *45*, 3414–3434. [CrossRef]
16. Demirci, U.B. Ammonia borane: An extensively studied, though not yet implemented, hydrogen carrier. *Energies* **2020**, *13*, 3071. [CrossRef]
17. Li, C.; Peng, P.; Zhou, D.W.; Wan, L. Research progress in LiBH$_4$ for hydrogen storage: A review. *Int. J. Hydrogen Energy* **2011**, *36*, 14512–14526. [CrossRef]
18. Luo, Y.; Sun, L.; Xu, F.; Liu, Z. Improved hydrogen storage of LiBH$_4$ and NH$_3$BH$_3$ by catalysts. *J. Mater. Chem. A* **2018**, *6*, 7293–7309. [CrossRef]

9. Liu, B.H.; Li, Z.P. A review: Hydrogen generation from borohydride hydrolysis reaction. *J. Power Sources* **2009**, *187*, 527–534. [CrossRef]
10. Abdelhamid, H.N. A review on hydrogen generation from the hydrolysis of sodium borohydride. *Int. J. Hydrogen Energy* **2021**, *46*, 726–765. [CrossRef]
11. Sun, Z.; Lu, X.; Nyahuma, F.M.; Yan, N.; Xiao, J.; Su, S.; Zhang, L. Enhancing Hydrogen Storage Properties of MgH_2 by Transition Metals and Carbon Materials: A Brief Review. *Front. Chem.* **2020**, *8*, 552. [CrossRef]
12. Perejón, A.; Sánchez-Jiménez, P.E.; Criado, J.M.; Pérez-Maqueda, L.A. Magnesium hydride for energy storage applications: The kinetics of dehydrogenation under different working conditions. *J. Alloys Compd.* **2016**, *681*, 571–579. [CrossRef]
13. Lamb, K.E.; Dolan, M.D.; Kennedy, D.F. Ammonia for hydrogen storage; A review of catalytic ammonia decomposition and hydrogen separation and purification. *Int. J. Hydrogen Energy* **2019**, *44*, 3580–3593. [CrossRef]
14. Aziz, M.; TriWijayanta, A.; Nandiyanto, A.B.D. Ammonia as effective hydrogen storage: A review on production, storage and utilization. *Energies* **2020**, *13*, 3062. [CrossRef]
15. Palo, D.R.; Dagle, R.A.; Holladay, J.D. Methanol Steam Reforming for Hydrogen Production. *Chem. Rev.* **2007**, *107*, 3992–4021. [CrossRef]
16. Onishi, N.; Laurenczy, G.; Beller, M.; Himeda, Y. Recent progress for reversible homogeneous catalytic hydrogen storage in formic acid and in methanol. *Coord. Chem. Rev.* **2018**, *373*, 317–332. [CrossRef]
17. Cheng, Y.; Wu, X.; Xu, H. Catalytic decomposition of hydrous hydrazine for hydrogen production. *Sustain. Energy Fuels* **2019**, *3*, 343–365. [CrossRef]
18. Zhou, L.; Luo, X.; Xu, L.; Wan, C.; Ye, M. Pt-ni nanoalloys for H_2 generation from hydrous hydrazine. *Catalysts* **2020**, *10*, 930. [CrossRef]
19. Joy, J.; Mathew, J.; George, S.C. Nanomaterials for photoelectrochemical water splitting—Review. *Int. J. Hydrogen Energy* **2018**, *43*, 4804–4817. [CrossRef]
20. Fajrina, N.; Tahir, M. A critical review in strategies to improve photocatalytic water splitting towards hydrogen production. *Int. J. Hydrogen Energy* **2019**, *44*, 540–577. [CrossRef]
21. Navlani-García, M.; Salinas-Torres, D.; Mori, K.; Kuwahara, Y.; Yamashita, H. Photocatalytic Approaches for Hydrogen Production via Formic Acid Decomposition. *Top. Curr. Chem.* **2019**, *377*, 27. [CrossRef]
22. Navlani-García, M.; Mori, K.; Salinas-Torres, D.; Kuwahara, Y.; Yamashita, H. New Approaches Toward the Hydrogen Production From Formic Acid Dehydrogenation Over Pd-Based Heterogeneous Catalysts. *Front. Mater.* **2019**, *6*, 44. [CrossRef]
23. Navlani-García, M.; Salinas-Torres, D.; Cazorla-Amorós, D. Hydrogen production from formic acid attained by bimetallic heterogeneous pdag catalytic systems. *Energies* **2019**, *12*, 4027. [CrossRef]
24. Srinivasan, S.; Demirocak, D.E.; Kaushik, A.; Sharma, M.; Chaudhary, G.R.; Hickman, N.; Stefanakos, E. Reversible hydrogen storage using nanocomposites. *Appl. Sci.* **2020**, *10*, 4618. [CrossRef]
25. Qin, G.; Cui, Q.; Yun, B.; Sun, L.; Du, A.; Sun, Q. High capacity and reversible hydrogen storage on two dimensional C_2N monolayer membrane. *Int. J. Hydrogen Energy* **2018**, *43*, 9895–9901. [CrossRef]
26. Cao, Z.; Ouyang, L.; Wang, H.; Liu, J.; Felderhoff, M.; Zhu, M. Reversible hydrogen storage in yttrium aluminum hydride. *J. Mater. Chem. A* **2017**, *5*, 6042–6046. [CrossRef]
27. Zhang, C. Hydrogen storage: Improving reversibility. *Nat. Energy* **2017**, *2*, 17064. [CrossRef]
28. Staubitz, A.; Robertson, A.P.M.; Manners, I. Ammonia-Borane and Related Compounds as Dihydrogen Sources. *Chem. Rev.* **2010**, *110*, 4079–4124. [CrossRef]
29. Kumar, R.; Karkamkar, A.; Bowden, M.; Autrey, T. Solid-state hydrogen rich boron–nitrogen compounds for energy storage. *Chem. Soc. Rev.* **2019**, *48*, 5350–5380. [CrossRef]
30. Marder, T.B. Will We Soon Be Fueling our Automobiles with Ammonia–Borane? *Angew. Chem. Int. Ed.* **2007**, *46*, 8116–8118. [CrossRef]
31. Shore, S.G.; Parry, R.W. The Crystalline Compound Ammonia-Borane, $^1H_3NBH_3$. *J. Am. Chem. Soc.* **1955**, *77*, 6084–6085. [CrossRef]
32. Valero-Pedraza, M.-J.; Cot, D.; Petit, E.; Aguey-Zinsou, K.-F.; Alauzun, J.G.; Demirci, U.B. Ammonia Borane Nanospheres for Hydrogen Storage. *ACS Appl. Nano Mater.* **2019**, *2*, 1129–1138. [CrossRef]
33. Stephens, F.H.; Pons, V.; Tom Baker, R. Ammonia–borane: The hydrogen source par excellence? *Dalt. Trans.* **2007**, 2613–2626. [CrossRef]
34. Zhan, W.W.; Zhu, Q.L.; Xu, Q. Dehydrogenation of Ammonia Borane by Metal Nanoparticle Catalysts. *ACS Catal.* **2016**, *6*, 6892–6905. [CrossRef]
35. Green, I.G.; Johnson, K.M.; Roberts, B.P. Homolytic reactions of ligated boranes. Part 13. An electron spin resonance study of radical addition to aminoboranes. *J. Chem. Soc. Perkin Trans.* **1989**, *2*, 1963–1972. [CrossRef]
36. Cheng, H.; Kamegawa, T.; Mori, K.; Yamashita, H. Surfactant-Free Nonaqueous Synthesis of Plasmonic Molybdenum Oxide Nanosheets with Enhanced Catalytic Activity for Hydrogen Generation from Ammonia Borane under Visible Light. *Angew. Chemie Int. Ed.* **2014**, *53*, 2910–2914. [CrossRef]
37. Verma, P.; Kuwahara, Y.; Mori, K.; Yamashita, H. Enhancement of Ag-based plasmonic photocatalysis in hydrogen production from ammonia borane by the assistance of single-site Ti-oxide moieties within a silica framework. *Chem. A Eur. J.* **2017**, *23*, 3616–3622. [CrossRef]

48. Wen, M.; Cui, Y.; Kuwahara, Y.; Mori, K.; Yamashita, H. Non-Noble-Metal Nanoparticle Supported on Metal-Organic Framework as an Efficient and Durable Catalyst for Promoting H_2 Production from Ammonia Borane under Visible Light Irradiation. *ACS Appl. Mater. Interfaces* **2016**, *8*, 21278–21284. [CrossRef]
49. Yin, H.; Kuwahara, Y.; Mori, K.; Yamashita, H. Plasmonic metal/$MoxW_{1-x}O_{3-y}$ for visible-light-enhanced H_2 production from ammonia borane. *J. Mater. Chem. A* **2018**, *6*, 10932–10938. [CrossRef]
50. Masuda, S.; Mori, K.; Sano, T.; Miyawaki, K.; Chiang, W.-H.; Yamashita, H. Simple Route for the Synthesis of Highly Active Bimetallic Nanoparticle Catalysts with Immiscible Ru and Ni Combination by utilizing a TiO_2 Support. *ChemCatChem* **2018**, *10*, 3526–3531. [CrossRef]
51. Fernández-Catalá, J.; Navlani-García, M.; Verma, P.; Berenguer-Murcia, Á.; Mori, K.; Kuwahara, Y.; Yamashita, H.; Cazorla-Amorós, D. Photocatalytically-driven H_2 production over Cu/TiO_2 catalysts decorated with multi-walled carbon nanotubes. *Catal. Today* **2021**, *364*, 182–189. [CrossRef]
52. Salinas-Torres, D.; Navlani-García, M.; Kuwahara, Y.; Mori, K.; Yamashita, H. Non-noble metal doped perovskite as a promising catalyst for ammonia borane dehydrogenation. *Catal. Today* **2019**, *351*, 6–11. [CrossRef]
53. García-Aguilar, J.; Navlani-García, M.; Berenguer-Murcia, Á.; Mori, K.; Kuwahara, Y.; Yamashita, H.; Cazorla-Amorós, D. Enhanced ammonia-borane decomposition by synergistic catalysis using CoPd nanoparticles supported on titano-silicates. *RSC Adv.* **2016**, *6*, 91768–91772. [CrossRef]
54. Liu, P.-H.; Wen, M.; Tan, C.-S.; Navlani-García, M.; Kuwahara, Y.; Mori, K.; Yamashita, H.; Chen, L.-J. Surface plasmon resonance enhancement of production of H2 from ammonia borane solution with tunable $Cu_{2-x}S$ nanowires decorated by Pd nanoparticles. *Nano Energy* **2017**, *31*, 57–63. [CrossRef]
55. Akbayrak, S.; Özkar, S. Ammonia Borane as Hydrogen Storage Materials. *Int. J. Hydrogen Energy* **2018**, *43*, 18592–18606. [CrossRef]
56. Akbayrak, S.; Tonbul, Y.; Özkar, S. Magnetically Separable Rh^0/Co_3O_4 Nanocatalyst Provides over a Million Turnovers in Hydrogen Release from Ammonia Borane. *ACS Sustain. Chem. Eng.* **2020**, *8*, 4216–4224. [CrossRef]
57. Akbayrak, S.; Çakmak, G.; Öztürk, T.; Özkar, S. Rhodium(0), Ruthenium(0) and Palladium(0) nanoparticles supported on carbon-coated iron: Magnetically isolable and reusable catalysts for hydrolytic dehydrogenation of ammonia borane. *Int. J. Hydrogen Energy* **2020**, *46*, 13548–13560. [CrossRef]
58. Tonbul, Y.; Akbayrak, S.; Özkar, S. Nanozirconia supported ruthenium(0) nanoparticles: Highly active and reusable catalyst in hydrolytic dehydrogenation of ammonia borane. *J. Colloid Interface Sci.* **2018**, *513*, 287–294. [CrossRef] [PubMed]
59. Jiang, H.-L.; Xu, Q. Catalytic hydrolysis of ammonia borane for chemical hydrogen storage. *Catal. Today* **2011**, *170*, 56–63. [CrossRef]
60. Umegaki, T.; Yabuuchi, K.; Yoshida, N.; Xu, Q.; Kojima, Y. In situ synthesized hollow spheres of a silica-ruthenium-nickel composite catalyst for the hydrolytic dehydrogenation of ammonia borane. *New J. Chem.* **2019**, *44*, 450–455. [CrossRef]
61. Chen, Y.; Yang, X.; Kitta, M.; Xu, Q. Monodispersed Pt nanoparticles on reduced graphene oxide by a non-noble metal sacrificial approach for hydrolytic dehydrogenation of ammonia borane. *Nano Res.* **2017**, *10*, 3811–3816. [CrossRef]
62. Li, J.; Zhu, Q.L.; Xu, Q. Non-noble bimetallic CuCo nanoparticles encapsulated in the pores of metal-organic frameworks: Synergetic catalysis in the hydrolysis of ammonia borane for hydrogen generation. *Catal. Sci. Technol.* **2015**, *5*, 525–530. [CrossRef]
63. Liu, M.; Zhou, L.; Luo, X.; Wan, C.; Xu, L. Recent advances in noble metal catalysts for hydrogen production from ammonia borane. *Catalysts* **2020**, *10*, 788. [CrossRef]
64. Chen, W.; Li, D.; Wang, Z.; Qian, G.; Sui, Z.; Duan, X.; Zhou, X.; Yeboah, I.; Chen, D. Reaction mechanism and kinetics for hydrolytic dehydrogenation of ammonia borane on a Pt/CNT catalyst. *AIChE J.* **2017**, *63*, 60–65. [CrossRef]
65. Xu, Q.; Chandra, M. Catalytic activities of non-noble metals for hydrogen generation from aqueous ammonia-borane at room temperature. *J. Power Sources* **2006**, *163*, 364–370. [CrossRef]
66. Peng, C.-Y.; Kang, L.; Cao, S.; Chen, Y.; Lin, Z.-S.; Fu, W.-F. Nanostructured Ni2P as a Robust Catalyst for the Hydrolytic Dehydrogenation of Ammonia-Borane. *Angew. Chem. Int. Ed.* **2015**, *54*, 15725–15729. [CrossRef]
67. Ma, H.; Na, C. Isokinetic temperature and size-controlled activation of ruthenium-catalyzed ammonia borane hydrolysis. *ACS Catal.* **2015**, *5*, 1726–1735. [CrossRef]
68. Cao, N.; Luo, W.; Cheng, G. One-step synthesis of graphene supported Ru nanoparticles as efficient catalysts for hydrolytic dehydrogenation of ammonia borane. *Int. J. Hydrogen Energy* **2013**, *38*, 11964–11972. [CrossRef]
69. Cui, Z.; Guo, Y.; Feng, Z.; Xu, D.; Ma, J. Ruthenium nanoparticles supported on nitrogen-doped porous carbon as a highly efficient catalyst for hydrogen evolution from ammonia borane. *New J. Chem.* **2019**, *43*, 4377–4384. [CrossRef]
70. Fan, G.; Liu, Q.; Tang, D.; Li, X.; Bi, J.; Gao, D. Nanodiamond supported Ru nanoparticles as an effective catalyst for hydrogen evolution from hydrolysis of ammonia borane. *Int. J. Hydrogen Energy* **2016**, *41*, 1542–1549. [CrossRef]
71. Fu, L.; Cai, L. Ru nanoparticles loaded on tannin immobilized collagen fibers for catalytic hydrolysis of ammonia borane. *Int. J. Hydrogen Energy* **2021**, *46*, 10749–10762. [CrossRef]
72. Akbayrak, S.; Özkar, S. Ruthenium(0) nanoparticles supported on multiwalled carbon nanotube as highly active catalyst for hydrogen generation from ammonia-borane. *ACS Appl. Mater. Interfaces* **2012**, *4*, 6302–6310. [CrossRef]
73. Cheng, W.; Zhao, X.; Luo, W.; Zhang, Y.; Wang, Y.; Fan, G. Bagasse-derived Carbon-supported Ru nanoparticles as Catalyst for Efficient Dehydrogenation of Ammonia Borane. *ChemNanoMat* **2020**, *6*, 1251–1259. [CrossRef]

74. Lu, R.; Xu, C.; Wang, Q.; Wang, Y.; Zhang, Y.; Gao, D.; Bi, J.; Fan, G. Ruthenium nanoclusters distributed on phosphorus-doped carbon derived from hypercrosslinked polymer networks for highly efficient hydrolysis of ammonia-borane. *Int. J. Hydrogen Energy* **2018**, *43*, 18253–18260. [CrossRef]
75. Liang, H.; Chen, G.; Desinan, S.; Rosei, R.; Rosei, F.; Ma, D. In situ facile synthesis of ruthenium nanocluster catalyst supported on carbon black for hydrogen generation from the hydrolysis of ammonia-borane. *Int. J. Hydrogen Energy* **2012**, *37*, 17921–17927. [CrossRef]
76. Zhong, F.; Wang, Q.; Xu, C.; Yang, Y.; Wang, Y.; Zhang, Y.; Gao, D.; Bi, J.; Fan, G. Ultrafine and highly dispersed Ru nanoparticles supported on nitrogen-doped carbon nanosheets: Efficient catalysts for ammonia borane hydrolysis. *Appl. Surf. Sci.* **2018**, *455*, 326–332. [CrossRef]
77. Fan, Y.; Li, X.; He, X.; Zeng, C.; Fan, G.; Liu, Q.; Tang, D. Effective hydrolysis of ammonia borane catalyzed by ruthenium nanoparticles immobilized on graphic carbon nitride. *Int. J. Hydrogen Energy* **2014**, *39*, 19982–19989. [CrossRef]
78. Du, C.; Ao, Q.; Cao, N.; Yang, L.; Luo, W.; Cheng, G. Facile synthesis of monodisperse ruthenium nanoparticles supported on graphene for hydrogen generation from hydrolysis of ammonia borane. *Int. J. Hydrogen Energy* **2015**, *40*, 6180–6187. [CrossRef]
79. Navlani-García, M.; Mori, K.; Nozaki, A.; Kuwahara, Y.; Yamashita, H. Highly efficient Ru/carbon catalysts prepared by pyrolysis of supported Ru complex towards the hydrogen production from ammonia borane. *Appl. Catal. A Gen.* **2016**, *527*, 45–52. [CrossRef]
80. Chu, H.; Li, N.; Qiu, X.; Wang, Y.; Qiu, S.; Zeng, J.-L.; Zou, Y.; Xu, F.; Sun, L. Poly(N-vinyl-2-pyrrolidone)-stabilized ruthenium supported on bamboo leaf-derived porous carbon for NH3BH3 hydrolysis. *Int. J. Hydrogen Energy* **2019**, *44*, 29255–29262. [CrossRef]
81. Chu, H.; Li, N.; Qiu, S.; Zou, Y.; Xiang, C.; Xu, F.; Sun, L. Ruthenium supported on nitrogen-doped porous carbon for catalytic hydrogen generation from NH3BH3 hydrolysis. *Int. J. Hydrogen Energy* **2019**, *44*, 1774–1781. [CrossRef]
82. Akbayrak, S.; Kaya, M.; Volkan, M.; Özkar, S. Ruthenium(0) nanoparticles supported on magnetic silica coated cobalt ferrite: Reusable catalyst in hydrogen generation from the hydrolysis of ammonia-borane. *J. Mol. Catal. A Chem.* **2014**, *394*, 253–261. [CrossRef]
83. Yao, Q.; Shi, W.; Feng, G.; Lu, Z.-H.; Zhang, X.; Tao, D.; Kong, D.; Chen, X. Ultrafine Ru nanoparticles embedded in SiO$_2$ nanospheres: Highly efficient catalysts for hydrolytic dehydrogenation of ammonia borane. *J. Power Sources* **2014**, *257*, 293–299. [CrossRef]
84. Deka, J.R.; Saikia, D.; Hsia, K.S.; Kao, H.M.; Yang, Y.C.; Chen, C.S. Ru nanoparticles embedded in cubic mesoporous silica SBA-1 as highly efficient catalysts for hydrogen generation from ammonia borane. *Catalysts* **2020**, *10*, 267. [CrossRef]
85. Yao, Q.; Lu, Z.-H.; Yang, K.; Chen, X.; Zhu, M. Ruthenium nanoparticles confined in SBA-15 as highly efficient catalyst for hydrolytic dehydrogenation of ammonia borane and hydrazine borane. *Sci. Rep.* **2015**, *5*, 15186. [CrossRef] [PubMed]
86. Sait Izgi, M.; Ece, M.Ş.; Kazici, H.Ç.; Şahin, Ö.; Onat, E. Hydrogen production by using Ru nanoparticle decorated with Fe$_3$O$_4$@SiO$_2$–NH$_2$ core-shell microspheres. *Int. J. Hydrogen Energy* **2020**, *45*, 30415–30430. [CrossRef]
87. Can, H.; Metin, Ö. A facile synthesis of nearly monodisperse ruthenium nanoparticles and their catalysis in the hydrolytic dehydrogenation of ammonia borane for chemical hydrogen storage. *Appl. Catal. B Environ.* **2012**, *125*, 304–310. [CrossRef]
88. Zhang, M.; Liu, L.; He, T.; Li, Z.; Wu, G.; Chen, P. Microporous Crystalline γ-Al$_2$O$_3$ Replicated from Microporous Covalent Triazine Framework and Its Application as Support for Catalytic Hydrolysis of Ammonia Borane. *Chem. Asian J.* **2017**, *12*, 470–475. [CrossRef] [PubMed]
89. Hu, M.; Wang, H.; Wang, Y.; Zhang, Y.; Wu, J.; Xu, B.; Gao, D.; Bi, J.; Fan, G. Alumina nanofiber-stabilized ruthenium nanoparticles: Highly efficient catalytic materials for hydrogen evolution from ammonia borane hydrolysis. *Int. J. Hydrogen Energy* **2017**, *42*, 24142–24149. [CrossRef]
90. Akbayrak, S.; Tanyıldızı, S.; Morkan, İ.; Özkar, S. Ruthenium(0) nanoparticles supported on nanotitania as highly active and reusable catalyst in hydrogen generation from the hydrolysis of ammonia borane. *Int. J. Hydrogen Energy* **2014**, *39*, 9628–9637. [CrossRef]
91. Kalkan, E.B.; Akbayrak, S.; Özkar, S. Ruthenium(0) nanoparticles supported on nanohafnia: A highly active and long-lived catalyst in hydrolytic dehydrogenation of ammonia borane. *Mol. Catal.* **2017**, *430*, 29–35. [CrossRef]
92. Wen, L.; Su, J.; Wu, X.; Cai, P.; Luo, W.; Cheng, G. Ruthenium supported on MIL-96: An efficient catalyst for hydrolytic dehydrogenation of ammonia borane for chemical hydrogen storage. *Int. J. Hydrogen Energy* **2014**, *39*, 17129–17135. [CrossRef]
93. Yang, K.; Zhou, L.; Yu, G.; Xiong, X.; Ye, M.; Li, Y.; Lu, D.; Pan, Y.; Chen, M.; Zhang, L.; et al. Ru nanoparticles supported on MIL-53(Cr, Al) as efficient catalysts for hydrogen generation from hydrolysis of ammonia borane. *Int. J. Hydrogen Energy* **2016**, *41*, 6300–6309. [CrossRef]
94. Zhang, S.; Zhou, L.; Chen, M. Amine-functionalized MIL-53(Al) with embedded ruthenium nanoparticles as a highly efficient catalyst for the hydrolytic dehydrogenation of ammonia-borane. *RSC Adv.* **2018**, *8*, 12282–12291. [CrossRef]
95. Cui, P.; Ren, H.; Zhu, G. Ruthenium Inlaying Porous Aromatic Framework for Hydrogen Generation from Ammonia Borane. *Front. Mater.* **2019**, *6*, 223. [CrossRef]
96. Zhao, W.; Wang, R.; Wang, Y.; Feng, J.; Li, C.; Chen, G. Effect of LDH composition on the catalytic activity of Ru/LDH for the hydrolytic dehydrogenation of ammonia borane. *Int. J. Hydrogen Energy* **2019**, *44*, 14820–14830. [CrossRef]

97. Qiu, X.; Liu, J.; Huang, P.; Qiu, S.; Weng, C.; Chu, H.; Zou, Y.; Xiang, C.; Xu, F.; Sun, L. Hydrolytic dehydrogenation of NH$_3$BH$_3$ catalyzed by ruthenium nanoparticles supported on magnesium–aluminum layered double-hydroxides. *RSC Adv.* **2020**, *10*, 9996–10005. [CrossRef]
98. Wu, Z.; Duan, Y.; Ge, S.; Yip, A.C.K.; Yang, F.; Li, Y.; Dou, T. Promoting hydrolysis of ammonia borane over multiwalled carbon nanotube-supported Ru catalysts via hydrogen spillover. *Catal. Commun.* **2017**, *91*, 10–15. [CrossRef]
99. Zhang, L.; Wang, Y.; Li, J.; Ren, X.; Lv, H.; Su, X.; Hu, Y.; Xu, D.; Liu, B. Ultrasmall Ru Nanoclusters on Nitrogen-Enriched Hierarchically Porous Carbon Support as Remarkably Active Catalysts for Hydrolysis of Ammonia Borane. *ChemCatChem* **2018**, *10*, 4910–4916. [CrossRef]
100. Li, Y.-T.; Zhang, S.-H.; Zheng, G.-P.; Liu, P.; Peng, Z.-K.; Zheng, X.-C. Ultrafine Ru nanoparticles anchored to porous g-C$_3$N$_4$ as efficient catalysts for ammonia borane hydrolysis. *Appl. Catal. A Gen.* **2020**, *595*, 117511. [CrossRef]
101. Navlani-garcía, M.; Verma, P.; Kuwahara, Y.; Kamegawa, T. Visible-light-enhanced catalytic activity of Ru nanoparticles over carbon modified g-C$_3$N$_4$. *J. Photochem. Photobiol. A* **2018**, *358*, 327–333. [CrossRef]
102. Chen, G.; Wang, R.; Zhao, W.; Kang, B.; Gao, D.; Li, C.; Lee, J.Y. Effect of Ru crystal phase on the catalytic activity of hydrolytic dehydrogenation of ammonia borane. *J. Power Sources* **2018**, *396*, 148–154. [CrossRef]
103. Konuş, N.; Karataş, Y.; Gulcan, M. In Situ Formed Ruthenium(0) Nanoparticles Supported on TiO$_2$ Catalyzed Hydrogen Generation from Aqueous Ammonia-Borane Solution at Room Temperature Under Air. *Synth. React. Inorganic Met. Nano-Metal Chem.* **2016**, *46*, 534–542. [CrossRef]
104. Wang, R.; Wang, Y.; Ren, M.; Sun, G.; Gao, D.; Chin Chong, Y.R.; Li, X.; Chen, G. Effect of ceria morphology on the catalytic activity of Ru/ceria for the dehydrogenation of ammonia borane. *Int. J. Hydrogen Energy* **2017**, *42*, 6757–6764. [CrossRef]
105. Nozaki, A.; Ueda, C.; Fujiwara, R.; Yamashita, A.; Yamamoto, H.; Morishita, M. Hydrogen Generation from Ammonia Borane over Ru/Nanoporous CeO$_2$ Catalysts Prepared from Amorphous Alloys. *Mater. Trans.* **2019**, *60*, 845–848. [CrossRef]
106. Dhakshinamoorthy, A.; Li, Z.; Garcia, H. Catalysis and photocatalysis by metal organic frameworks. *Chem. Soc. Rev.* **2018**, *47*, 8134–8172. [CrossRef]
107. Pascanu, V.; González Miera, G.; Inge, A.K.; Martín-Matute, B. Metal–Organic Frameworks as Catalysts for Organic Synthesis: A Critical Perspective. *J. Am. Chem. Soc.* **2019**, *141*, 7223–7234. [CrossRef]
108. Ranocchiari, M.; Bokhoven, J.A. van Catalysis by metal–organic frameworks: Fundamentals and opportunities. *Phys. Chem. Chem. Phys.* **2011**, *13*, 6388–6396. [CrossRef]
109. Chandra, M.; Xu, Q. A high-performance hydrogen generation system: Transition metal-catalyzed dissociation and hydrolysis of ammonia-borane. *J. Power Sources* **2006**, *156*, 190–194. [CrossRef]
110. Rachiero, G.P.; Demirci, U.B.; Miele, P. Bimetallic RuCo and RuCu catalysts supported on γ-Al$_2$O$_3$. A comparative study of their activity in hydrolysis of ammonia-borane. *Int. J. Hydrogen Energy* **2011**, *36*, 7051–7065. [CrossRef]
111. Valero-Pedraza, M.-J.; Alligier, D.; Petit, E.; Cot, D.; Granier, D.; Adil, K.; Yot, P.G.; Demirci, U.B. Diammonium tetraborate dihydrate as hydrolytic by-product of ammonia borane in aqueous alkaline conditions. *Int. J. Hydrogen Energy* **2020**, *45*, 9927–9935. [CrossRef]
112. Yadav, M.; Xu, Q. Liquid-phase chemical hydrogen storage materials. *Energy Environ. Sci.* **2012**, *5*, 9698–9725. [CrossRef]
113. Lang, C.; Jia, Y.; Yao, X. Recent advances in liquid-phase chemical hydrogen storage. *Energy Storage Mater.* **2020**, *26*, 290–312. [CrossRef]
114. Liu, C.-H.; Wu, Y.-C.; Chou, C.-C.; Chen, B.-H.; Hsueh, C.-L.; Ku, J.-R.; Tsau, F. Hydrogen generated from hydrolysis of ammonia borane using cobalt and ruthenium based catalysts. *Int. J. Hydrogen Energy* **2012**, *37*, 2950–2959. [CrossRef]
115. Ramachandran, P.V.; Gagare, P.D. Preparation of ammonia borane in high yield and purity, methanolysis, and regeneration. *Inorg. Chem.* **2007**, *46*, 7810–7817. [CrossRef]
116. Sutton, A.D.; Burrell, A.K.; Dixon, D.A.; Garner III, E.B.; Gordon, J.C.; Nakagawa, T.; Ott, K.C.; Robinson, J.P.; Vasiliu, M. Regeneration of ammonia borane spent fuel by direct reaction with hydrazine and liquid ammonia. *Science* **2011**, *331*, 1426–1429. [CrossRef]
117. Hajari, A.; Roy, B.; Kumar, V.; Bishnoi, A.; Sharma, P. Regeneration of Supported Ammonia Borane to Achieve Higher Yield. *ChemistrySelect* **2021**, *6*, 1276–1282. [CrossRef]

Article

Pd Catalysts Supported on Bamboo-Like Nitrogen-Doped Carbon Nanotubes for Hydrogen Production

Arina N. Suboch and Olga Y. Podyacheva *

Boreskov Institute of Catalysis SB RAS, 630090 Novosibirsk, Russia; arina@catalysis.ru
* Correspondence: pod@catalysis.ru

Abstract: Bamboo-like nitrogen-doped carbon nanotubes (N-CNTs) were used to synthesize supported palladium catalysts (0.2–2 wt.%) for hydrogen production via gas phase formic acid decomposition. The beneficial role of nitrogen centers of N-CNTs in the formation of active isolated palladium ions and dispersed palladium nanoparticles was demonstrated. It was shown that although the surface layers of N-CNTs are enriched with graphitic nitrogen, palladium first interacts with accessible pyridinic centers of N-CNTs to form stable isolated palladium ions. The activity of Pd/N-CNTs catalysts is determined by the ionic capacity of N-CNTs and dispersion of metallic nanoparticles stabilized on the nitrogen centers. The maximum activity was observed for the 0.2% Pd/N-CNTs catalyst consisting of isolated palladium ions. A ten-fold increase in the concentration of supported palladium increased the contribution of metallic nanoparticles with a mean size of 1.3 nm and decreased the reaction rate by only a factor of 1.4.

Keywords: hydrogen; formic acid; palladium; nitrogen; carbon nanotubes

---

Citation: Suboch, A.N.; Podyacheva, O.Y. Pd Catalysts Supported on Bamboo-Like Nitrogen-Doped Carbon Nanotubes for Hydrogen Production. Energies 2021, 14, 1501. https://doi.org/10.3390/en14051501

Academic Editor: Javier Fermoso

Received: 18 February 2021
Accepted: 3 March 2021
Published: 9 March 2021

Publisher's Note: MDPI stays neutral with regard to jurisdictional claims in published maps and institutional affiliations.

Copyright: © 2021 by the authors. Licensee MDPI, Basel, Switzerland. This article is an open access article distributed under the terms and conditions of the Creative Commons Attribution (CC BY) license (https://creativecommons.org/licenses/by/4.0/).

1. Introduction

Rational use of fossil fuel resources is a necessary condition for the long-term stable development of society. In this context, the development of efficient processes for energy production from renewable sources is a topical problem that can be solved using various fundamental studies, particularly in the field of material science. Hydrogen production from formic acid (FA) is a way to involve renewable biomass sources into the energy cycle [1,2]. The FA decomposition reaction proceeds on mono-, bi- and trimetallic catalysts deposited on oxide or carbon supports [3,4]. Activity and selectivity of the catalysts towards hydrogen production are determined by the nature of their active component, its dispersion, and electronic state of metals. The application of nitrogen-doped carbon nanomaterials (N-CNMs) as the catalyst supports makes it possible to perform the tailored synthesis of metal supported catalysts with desired properties [5].

It is known that nitrogen in N-CNMs having different structures occupies standard positions: pyridinic (N_{Py}), pyrrolic (N_{Pyr}), graphitic (N_Q), and oxidized (N_{Ox}). According to data in the literature, supported metals can stabilize on different nitrogen centers. For example, Pt stabilization on N_Q centers of nitrogen-doped carbon nanotubes (N-CNTs) leads to the formation of electron-rich Pt nanoparticles, which are active towards the hydrogenation of nitroarenes [6] and aerobic oxidation of glycerol [7]. Ombaka et al. [8] reported the formation of Pd^{2+} species on the N_{Pyr} centers of N-CNTs, which are active towards the hydrogenation of nitrobenzophenone [8]. However, a substantial body of data indicates the involvement of N_{Py} centers in the stabilization of electron-deficient metals on the surface of various N-CNMs, which show the increased activity and selectivity in hydrogenation reactions or electrochemical reduction of oxygen [9–12]. The positive role of the electron-deficient metals deposited on ordered or disordered N-CNMs was also demonstrated for FA decomposition in a liquid or gaseous medium [4,13–16]. Additionally, for platinum deposited on nitrogen-doped carbon nanofibers (N-CNFs) and N-CNTs, it was shown that the catalytic properties of metal in this reaction are determined by its locally one-type interaction with N_{Py} centers [16]. Among various metals that are active towards FA decomposition, platinum and palladium are the most active catalysts for this reaction [3,17]. It should be noted that the use of N-CNMs as the supports opens new possibilities for enhancing the efficiency of catalysts via the stabilization of highly active and selective atomic metal species [18,19]. Among the wide variety of N-CNMs, bamboo-like N-CNTs are characterized by a unique combination of the graphene layer curvature and the availability of all nitrogen centers. At the same time, data on the use of these materials as a support for the catalysts of the FA decomposition reaction are quite scarce [18].

This paper is devoted to investigating the formation of the Pd catalyst on bamboo-like N-CNTs for the gas phase formic acid decomposition reaction. A detailed investigation of the properties of N-CNTs, variation of palladium concentration, and application of various research methods allowed us to describe the interaction of palladium with the surface of N-CNTs. Different active palladium species were shown to form as isolated ions and metallic nanoparticles with a size of 1.3 nm, the ratio of which were determined by the affinity of palladium for N_{Py} centers and the palladium concentration in catalysts.

2. Materials and Methods

Nitrogen-doped and nitrogen-free carbon nanotubes (N-CNTs and CNTs) were synthesized by the standard CVD route [20]. An ethylene–ammonia mixture or ethylene were decomposed on the Fe-containing catalyst at 650–700 °C. The growth catalyst was removed by the boiling of N-CNTs and CNTs in 2 M hydrochloric acid. Nevertheless, X-ray photoelectron spectroscopy (XPS) revealed the presence of iron in the washed samples with the content up to approximately 0.5 at.%. Previously, we have demonstrated the low intrinsic activity of washed N-CNTs and CNTs in FA decomposition reactions compared to metal supported catalysts [16]. Palladium was deposited on carbons, which were pre-dried in Ar at 170 °C, by incipient wetness impregnation from a Pd acetate–acetone solution. The samples were dried in air at 105 °C for 8 h and then reduced in an H_2/Ar flow at 200 °C for 1 h.

Nitrogen adsorption measurements were carried out at 77 K by the use of ASAP-2400. Prior to measurements, the samples were evacuated at 200 °C for 24 h. Acetone capacity of the N-CNTs and CNTs was measured by the standard incipient wetness impregnation method. Acetone was added dropwise under stirring to the sample dried in Ar at 170 °C until the pores of the material were completely filled. The volume of the absorbed acetone (cm^3/g) was determined by weighing. CO chemisorption on the catalysts was measured at room temperature by a pulse technique. Before the experiments, the catalysts were reduced in hydrogen flow at 200 °C. X-ray diffraction patterns were recorded using an HZG-4 (Zeiss, Germany) diffractometer with monochromatic CuK_α radiation. Electron microscopy investigation was performed using JEM-2200 FS (JEOL Ltd., Japan) and Titan 60-300 (FEI, Netherlands) electron microscopes. Raman spectroscopy with an excitation wavelength of 632.8 nm was performed using a HORIBA LabRAM HR800 (HORIBA Jobin Yvon Raman Division, France) Raman spectrometer. XPS study of N-CNTs was performed on an ES-300 (KRATOS Analytical, UK) photoelectron spectrometer with an $AlK\alpha$ source ($h\nu$ = 1486.6 eV). The distribution of nitrogen in N-CNTs was studied by synchrotron-based photoelectron spectroscopy at 500 and 800 eV photon energy [21]. Pd catalysts were examined using a VG ESCALAB HP (ThermoScientific, UK) instrument with an $AlK\alpha$ source ($h\nu$ = 1486.6 eV). In situ reduction of the catalysts under 1000 Pa H_2 in a preparation chamber of the XPS unit was performed at 300 °C for 1 h [22]. After the pretreatment, the sample was transferred to an analytical chamber without contact with air.

Activity of the catalysts in the decomposition of FA (5 vol. % HCOOH/He) was measured using a flow setup. A 20 mg catalyst sample was uniformly mixed with 0.5 cm^3 of quartz sand (grain size of 0.25–0.5 mm). The reaction mixture preheated to 100 °C in a special box was fed into the reactor at a rate of 20 cm^3/min. The experiments were carried out in the temperature-programmed mode at a ramp rate of 2°/min with chromatographic analysis of the gas mixture. Before measuring the activity, the catalyst was reduced in a 10% H_2/He mixture at 200 °C for 1 h. The apparent turnover frequency (TOF) was calculated as a ratio of the reaction rate measured at low (<20%) conversions to the total number of palladium atoms in the catalysts.

3. Results

3.1. Properties of N-CNTs

Nitrogen-doped carbon nanotubes were synthesized by the CVD method, which is widely applied to obtain N-CNTs [23–26]. As seen in Table 1, by increasing the ammonia concentration in the C_2H_4–NH_3 reaction mixture (25, 40, 60 and 75% NH_3), one can monotonically increase the degree of nitrogen doping of nanotubes from 1.8 to 6.6 at.% with a simultaneous decrease in the yield of N-CNTs from 38 to 5 g C/g of catalyst. Such an effect of N/C ratio on the nitrogen content in N-CNTs was described by other authors for different reaction mixtures [24,26,27]. In turn, a decrease in the yield of N-CNTs is related to deactivation of the catalyst due to its encapsulation (Figure 1), which has been reported by many authors for the CVD synthesis of N-CNTs and other carbon materials [25,26,28–30].

Table 1. Properties of nitrogen-doped carbon nanotubes (N-CNTs): N/C and N_{Py}/N_Q ratios according to XPS data, I_D/I_G and I_{2D}/I_D ratios according to Raman spectroscopy, S_{BET} values and product yield.

Sample-NH_3 Concentration (%)	N/C at.%	N_{Py}/N_Q	I_D/I_G	I_{2D}/I_D	S_{BET}, (m^2/g)	Yield, (g C/g of Catalyst)
CNTs	-	-	2.4	0.28	155	38
N-CNTs-25	1.8	0.5	2.7	0.15	148	30
N-CNTs-40	2.8	0.7	2.8	0.15	160	22
N-CNTs-60	4.4	0.9	3.0	0.13	158	14
N-CNTs-75	6.6	1.2	3.7	0.12	151	5

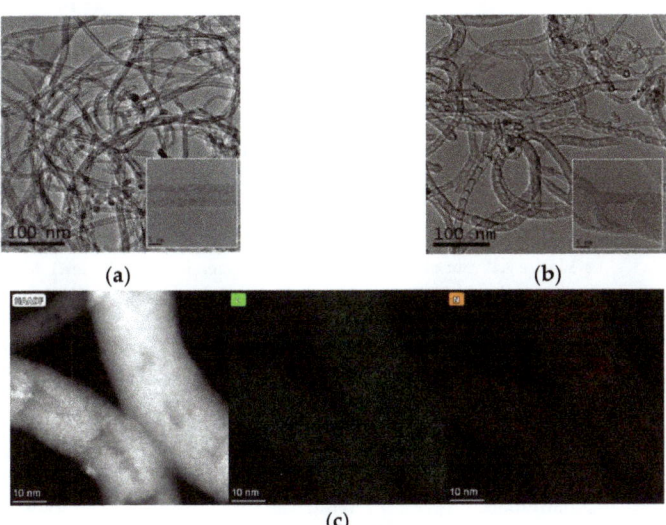

Figure 1. (a) TEM images of CNTs; (b) 4.4 at.% N-CNTs; and (c) HAADF-STEM image, color overlay of the corresponding EDX maps of carbon and nitrogen presented in background-corrected intensities.

TEM micrographs of the 4.4 at.% N-CNTs as an example and CNTs are displayed in Figure 1. Ethylene decomposition is accompanied by the predominant formation of multiwalled carbon nanotubes. In their turn, N-CNTs have a distinct bamboo-like structure, which is known to form due to the curvature of nitrogen-containing graphene layers [31,32].

According to STEM-HAADF images with elemental mapping, nitrogen is uniformly distributed in N-CNTs.

XPS revealed (Figure 2) that nitrogen in N-CNTs is in pyridinic (N_{Py}, ~398 eV), pyrrolic (N_{Pyr}, ~400 eV), graphitic (N_Q, ~401 eV), oxidized (N_{Ox}, ~403 eV) and molecular (N_{N2}, ~405 eV) states [24,26,33]. The presence of molecular nitrogen encapsulated between graphene layers or inside channels is characteristic of bamboo-like N-CNTs [26,34]. An increase in the total nitrogen content in N-CNTs is accompanied by an increase in the content of all structural species. However, this changes the predominant species and the ratio of nitrogen species. In the N-CNTs with the minimum doping degree, the main contribution is made by graphitic nitrogen, and the N_{Py}/N_Q ratio is equal to 0.5, Table 1. As the total nitrogen content is increased, the contribution of pyridinic nitrogen grows; as a result, the N_{Py}/N_Q value achieves 1.2 in the case of 6.6 at.% N-CNTs.

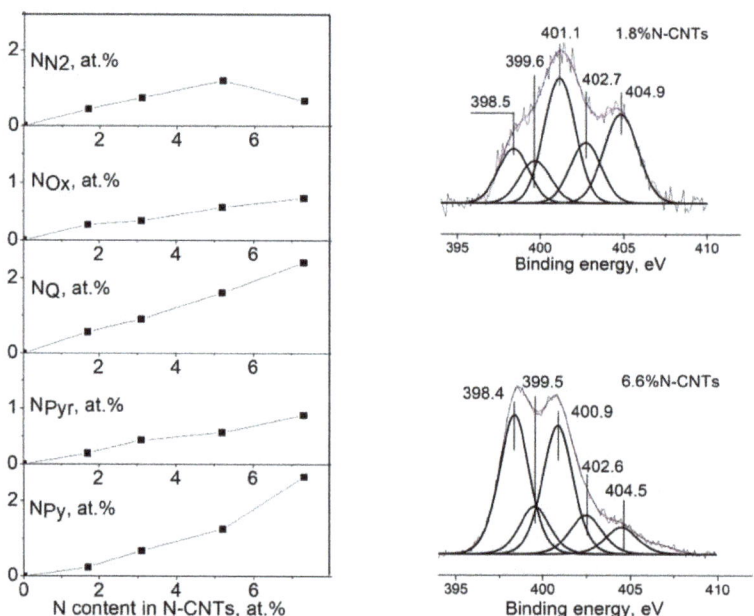

Figure 2. Input of different types of nitrogen in N-CNTs (**left**); N1s XPS spectra of 1.8% at.% N-CNTs and 6.6% at.% N-CNTs (**right**).

It is known that nitrogen incorporation in the pyridinic positions of graphene layers is accompanied by the formation of carbon vacancies, i.e., by an increase in structural defectivity. Indeed, the formation of such fragments was observed in TEM images of N-CNTs obtained at atomic resolution [23]. The formation of such structural elements was confirmed by a growth of the I_D/I_G ratio with a simultaneous decrease in the I_{2D}/I_D ratio revealed by Raman spectroscopy (Table 1). The I_D/I_G ratio reflects defectivity of carbon nanomaterials [35,36], while the I_{2D}/I_D ratio is used to analyze the size of the defect-free graphene blocks [37,38] (Figure 3). In our earlier work, the XRD study of N-CNTs and their structural simulation gave grounds to suppose the formation of the fragments similar to the structural elements of the g-C_3N_4 layer comprising the ordered carbon vacancies and pyridinic nitrogen atoms [20]; the estimated size of such a fragment of the layer is not smaller than 5 Å.

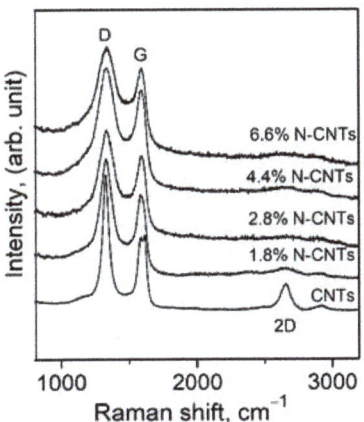

Figure 3. Raman spectra of CNTs and N-CNTs.

However, it should be noted that the obtained XPS data do not provide information on the surface state of N-CNTs with respect to different nitrogen species, which is an important factor when these materials are employed as the catalyst supports. In this context, N-CNTs were additionally studied by synchrotron-based XPS with variation of the analysis depth using radiation energies of 500 and 800 eV. The data obtained were compared with the results of standard XPS study performed at 1486.6 eV. The radiation energies of 500, 800 and 1486.6 eV allowed us to acquire data at the analysis depths of ca. 1.6, ca. 3 and ca. 6 nm, respectively (Figure 4). According to TEM data displayed in Figure 1b, the mean wall thickness of N-CNTs is ca. 5 nm. Hence, when the depth of analysis is varied from 1.6 to 6 nm, the acquired data could provide information on the nitrogen in external/internal graphene layers, and partially in internal arches.

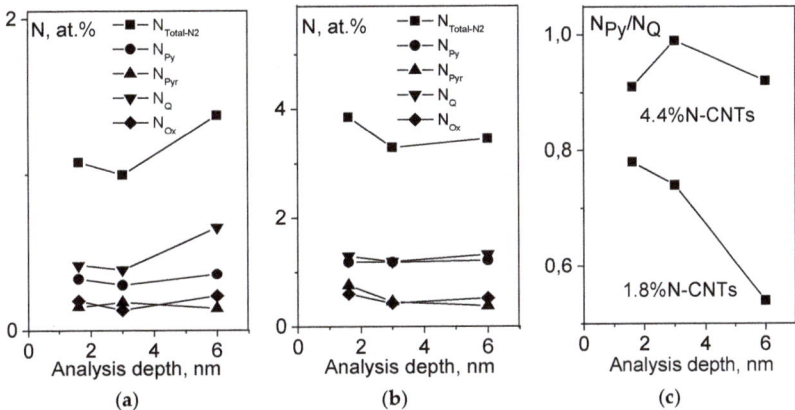

Figure 4. Content of nitrogen species in (**a**) 1.8 at.% N-CNTs and in (**b**) 4.4 at.% N-CNTs and (**c**) N_{Py}/N_Q ratio in N-CNTs in dependence on the analysis depth

One can see that in both samples, nitrogen is incorporated in all structural positions of external/internal graphene layers and internal arches. The total nitrogen concentration profile along the analysis depth coincides to a greater extent with the profile of different nitrogen species, which is determined most likely by the catalytic growth mechanism of N-CNTs on the Fe catalyst with the equiprobable incorporation of nitrogen in different positions [24]. As seen in Figure 4c, when the entire tube wall (ca. 6 nm) or only its external/subsurface layers (ca. 1.5 nm) are analyzed, N_{Py}/N_Q, ratios for 4.4 at.% N-CNTs virtually do not differ. In the case of 1.8 at.% N-CNTs, as the depth of analysis is decreased, the N_{Py}/N_Q ratio increases by a factor of ca. 1.6. Nevertheless, our study of N-CNTs with the minimum depth of analysis allows a conclusion that external layers of N-CNTs, irrespective of the doping degree, are slightly enriched with graphitic nitrogen.

To choose a method of palladium deposition on N-CNTs, their textural and adsorption properties were investigated. As seen in Table 1, nitrogen doping of CNTs does not produce changes in the textural characteristics of carbon nanotubes. Indeed, their specific surface area changes in a narrow range of 148–160 m^2/g, and pore size distribution curves for all the samples are similar and have distinct maxima in the regions of 3–4 nm and 25–40 nm (Figure 5). Fenelonov et al. [39] proposed that pores with the size of 3–4 nm are formed due to the curvature of graphene layers, whereas larger pores emerge due to the entangled weave of N-CNTs leading to the formation of multiple voids.

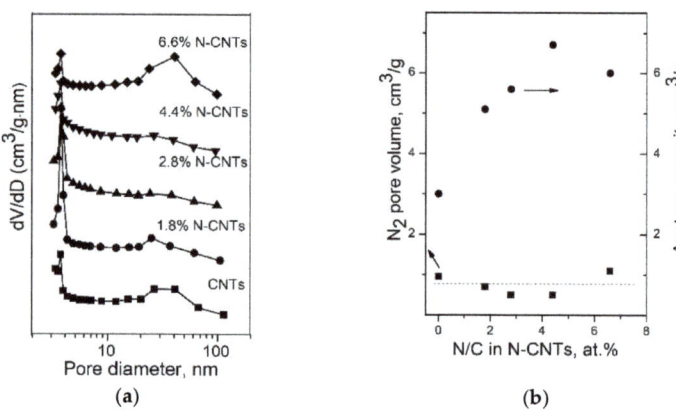

Figure 5. (a) Pore size distribution curves and (b) N_2 pore volume and acetone capacity of the CNTs and N-CNTs.

However, an increase in the nitrogen content in N-CNTs is accompanied by a considerable growth of their capacity towards acetone, from 3 to ~ 6–7 cm^3/g, which was used as a solvent in the deposition of palladium catalyst (Figure 5b). This suggests that such a high acetone capacity of N-CNTs is achieved due to the bulk filling of inner channels of the tubes in spite of multiple regular internal arches (Figure 1b). Note that the values of acetone capacity of fresh and treated in 2 M HCl carbons are the same. In turn, a good correlation between nitrogen content in N-CNTs and N_{Py}/N_Q, I_D/I_G and I_{2D}/D ratios was found (Table 1). Therefore, it can be supposed that the presence of large defects (not smaller than 5 Å) in graphene layers of the synthesized N-CNTs, which are represented by the ordered carbon vacancies surrounded with pyridinic nitrogen atoms [20], improves their wettability and makes the capillary effect more pronounced. Kumar et al. [40] calculated isotherms of water adsorption on disordered N–C materials and attributed the anomalously high adsorption to the implementation of a specific mechanism, according to which 1D, 2D and 3D nucleation, accompanied by the formation of solvent clusters which resembles the growth of crystals, was followed by bulk condensation. The filling of the inner channels of N-CNTs may indicate that during their deposition on N-CNTs by wet chemistry methods, metals can be anchored not only on the external surface but also in the inner channels of nanotubes [41]. On the other hand, Hao et al. [42] conclude that high hydrophilicity of N-doped carbon nanomaterials ensures a high dispersion of supported metallic catalysts, thus increasing their activity.

3.2. Properties of Pd/N-CNTs Catalysts

A series of Pd/N-CNTs catalysts (0.2, 0.5, 1 and 2 wt.% of Pd) were synthesized using the N-CNTs that were obtained by decomposition of the 40% C_2H_4–60% NH_3 mixture. The choice of support was based on the optimal combination of the nitrogen content in carbon nanotubes and the yield of N-CNTs (Table 1). To study the interaction of palladium with the N-CNTs surface, the concentration of palladium was varied from 0.2 to 2 wt.%. N-free catalysts 0.2% Pd/CNTs and 2% Pd/CNTs were synthesized for comparison purposes. The revealed high affinity of N-CNTs for acetone, which was used in our study as a solvent of palladium salt, explains the uniform distribution of palladium in the catalysts and the absence of agglomerates (Figure 6). In addition, as seen in Figure 6b, it cannot be ruled out that individual palladium particles are located in the inner channels of nanotubes. In the X-ray diffraction patterns, a weak intensity peak related to Pd 111 was recorded only for the catalysts with a maximum metal content (2 wt.%; see Figure 7). The X-ray diffraction patterns of all other catalysts show only the maxima related to the carbon phase.

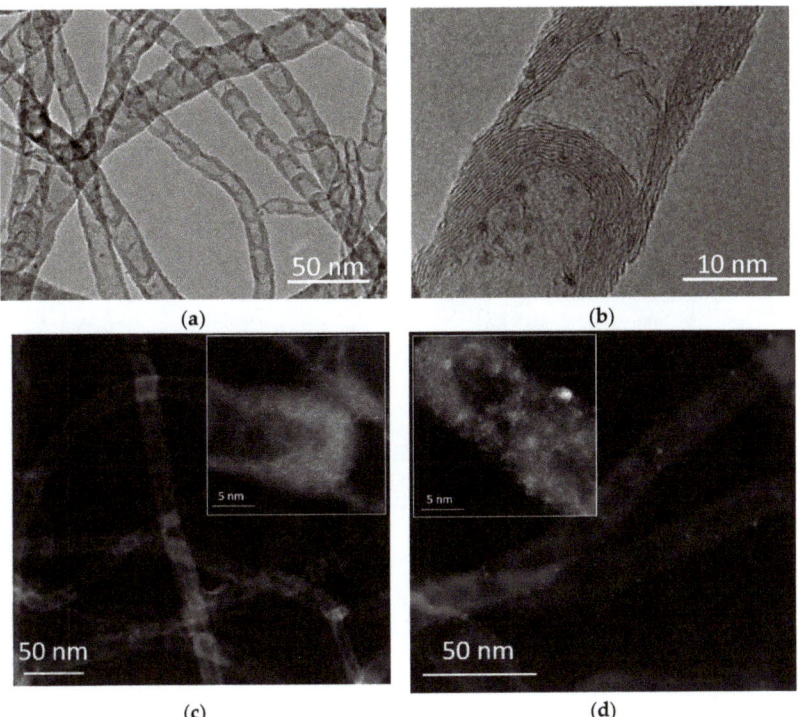

Figure 6. (a,b) TEM images of 2% Pd/N-CNTs; HAADF-STEM images of (c) 0.2% Pd/N-CNTs and (d) 2%Pd/N-CNTs

TOF values for FA decomposition on nitrogen-doped and nitrogen-free catalysts are shown in Table 2 and Figure 8. In the case of N-free catalysts, a ten-fold increase in palladium content was accompanied by a decrease in apparent TOF from 180 to 72 h^{-1}, with a simultaneous decrease in CO/Pd ratio from 33.5 to 22.1%, according to the CO chemisorption data. Indeed, the mean particle size for 0.2% Pd/CNTs was 1.2 nm, whereas for 2% Pd/CNTs it was 2.3 nm. According to the XPS, palladium was in the metallic state and the binding energy of Pd 3d was 335.6 eV (Figure 9). This value is typical of the dispersed metallic palladium particles [43,44]. The TOF values of the 0.2% Pd/CNTs and 2% Pd/CNTs catalysts calculated as a ratio of the reaction rate to the number of surface palladium atoms were 193 and 147 h^{-1}, respectively. The observed difference in TOF values may be due to different palladium crystallinity or its different localization in the tubes (outer surface and inner channels). Selectivity of the catalysts towards hydrogen formation did not exceed 92%.

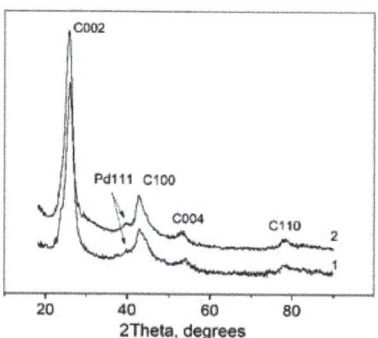

Figure 7. XRD patterns of 2% Pd/N-CNTs (1) and 2% Pd/CNTs (2).

Table 2. Some characteristics of Pd supported catalysts: mean Pd size determined by TEM, TOF at 125 °C and selectivity at 25% formic acid (FA) conversion.

Catalyst	Pd Size (nm)	TOF (h^{-1})	S (%)
0.2% Pd/N-CNTs	not detectable	324	97
0.5% Pd/N-CNTs	not detectable	252	97
1% Pd/N-CNTs	1.3	297	98
2% Pd/N-CNTs	1.4	225	98
0.2% Pd/CNTs	1.2	180	92
2% Pd/CNTs	2.3	72	92

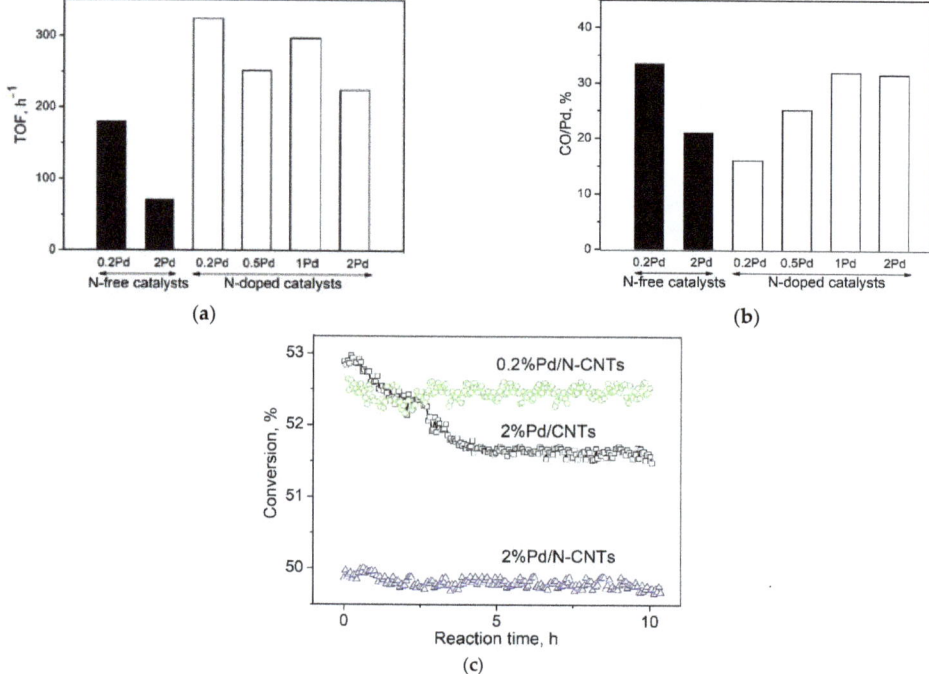

Figure 8. (a) TOF values for FA decomposition at 125 °C; (b) CO/Pd ratio for nitrogen-free and nitrogen-doped catalysts with different Pd loading and stability test of the 0.2% Pd/N-CNTs, 2% Pd/N-CNTs and (c) 2% Pd/CNTs at ~50% FA conversion in the course of the reaction.

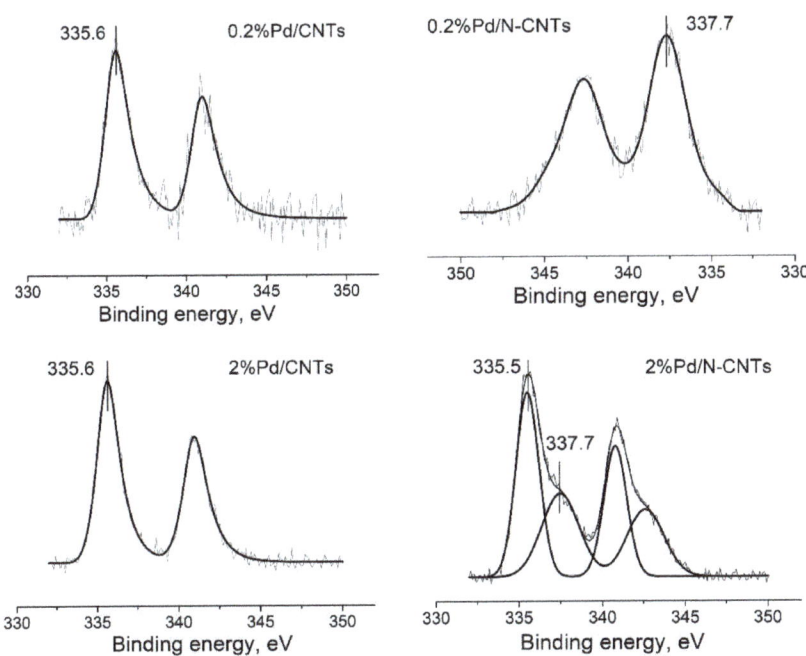

Figure 9. Pd 3d XPS spectra of nitrogen-free (0.2% Pd/CNTs and 2% Pd/CNTs) and nitrogen-doped (0.2% Pd/N-CNTs and 2% Pd/N-CNTs) catalysts.

All the nitrogen-doped catalysts, irrespective of palladium content, demonstrated the increased activity and selectivity (97–98%) in comparison with 0.2–2% Pd/CNTs catalysts. TOF values varied in a narrow range of 225–324 h^{-1}; the maximum activity was observed for the 0.2% Pd/N-CNT catalyst. The dependence of CO/Pd on palladium content for this series of catalysts had a complicated shape. A decrease in palladium content from 2 to 1% was not accompanied by changes in CO/Pd, and a further decrease in Pd concentration to 0.2% produced a sharp drop in CO/Pd from 32 to 16.1%. The observed dependence of CO/Pd on palladium content did not correlate with the TEM data, according to which mean particle size for the catalysts with 2% Pd and 1% Pd was equal to 1.4 and 1.3 nm, respectively (Table 2), whereas for 0.5% Pd and 0.2% Pd it could not be determined owing to their small sizes. An HAADF-STEM study of 0.2% Pd/N-CNTs showed that the catalyst consisted predominantly of isolated palladium atoms (Figure 6). The 2% Pd/N-CNT catalyst contained not only nanoparticles but also a considerable amount of single palladium atoms. As seen in Figure 9, the binding energy of palladium in 0.2% Pd/N-CNTs was 337.7 eV, which may correspond to both the PdO [45] and the ionic palladium stabilized on pyridinic centers of N-CNTs [46]. However, oxidation of the samples was completely excluded in XPS studies; therefore, Pd 3d BE 337.7 eV can reliably be assigned to Pd^{2+}-N_{Py} species. Some studies have demonstrated that Pd^{2+}-N_{Py} species, in distinction to Pd^0, weakly chemisorb CO [12,46]. In the case of 2% Pd/N-CNTs, palladium is in the highly dispersed Pd^0 (BE Pd 3d 335.5 eV) and Pd^{2+}-N_{Py} (BE Pd 3d 337.7 eV) states.

4. Discussion

Pd/N-CNTs catalysts have been synthesized and examined in the gas-phase decomposition of FA for hydrogen production. To describe the interaction of palladium with N-CNTs, palladium concentration was varied from 0.2 to 2 wt.%, and a comparison with N-free catalysts was made. XPS experiments with variation of the analysis depth demonstrated that nitrogen on the external graphene surface of N-CNTs was in all structural states.

Irrespective of the doping degree, the predominance of graphitic nitrogen was observed. However, according to XPS and HAADF-STEM studies of the 0.2% Pd/N-CNT catalyst, palladium starts interacting with pyridinic nitrogen centers of N-CNTs to produce isolated ions. Indeed, theoretical calculations show that palladium preferentially interacts with the N_{Py} centers of N-CNTs instead of the N_Q centers to form a positive +0.5 e charge [47]. The binding energy of Pd 3d in 0.2% Pd/N-CNTs is 337.7 eV, which coincides well with the BE Pd 3d values for Pd/N-CNTs catalysts reported in literature [47,48]. In these studies, DFT calculations were used to propose the models of palladium anchoring on both the single N_{Py} center and $3N_{Py}$-V (where V is the vacancy). It seems interesting that close BE Pd 3d values of 336.9–338.5 eV were also obtained in the case of palladium deposition on g-C_3N_4 or CN material derived from a metal–organic framework [12,19]. It was supposed that stabilization of isolated palladium ions occurs with the participation of four nitrogen centers ($2N_{Py}$ and $2N_{Pyr}$) [19] or $6N_{Py}$, which forms structural elements of the g-C_3N_4 graphene layer [12]. It should be noted that in the 0.2% Pd/N-CNTs catalyst, the N_{Py}/Pd ratio is 25, which makes it possible to implement any of the proposed models, including Pd anchoring on fragments of the layer, which are similar to g-C_3N_4. The possibility of the formation of such fragments in graphene layers of N-CNTs was demonstrated in our earlier study [20].

As the content of supported palladium is increased, this most likely leads to the filling of accessible N_{Py} centers; as a result, metallic nanoparticles start to form (Figure 10). This assumption is supported by the obtained dependence of CO/Pd on palladium content in the catalysts. It reflects the occurrence of competitive processes: an increase in CO chemisorption owing to the growing dispersion of metallic nanoparticles, and a decrease in CO chemisorption due to formation of ionic palladium. It should be noted that the involvement of N_Q centers in the stabilization of metallic nanoparticles, which are more dispersed than in N-free catalysts, cannot be ruled out; their involvement was derived, for example, by Shi et al. [6] from a 0.2 eV shift of the corresponding maximum in the N1s spectrum towards higher binding energies. We did not observe any changes in the N1s spectrum of Pd/N-CNTs catalysts in comparison with N-CNTs. However, a 0.1 eV shift of BE Pd 3d for a more dispersed 2% Pd/N-CNTs (1.4 nm) towards lower binding energies in comparison with 2% Pd/CNTs (2.3 nm) provides the basis to suppose such a possibility. As a result, the involvement of nitrogen centers in the stabilization of palladium decreases the TOF value only by a factor of 1.4 at a ten-fold increase in palladium concentration. For N-free catalysts, the TOF value decreases by a factor of 2.5.

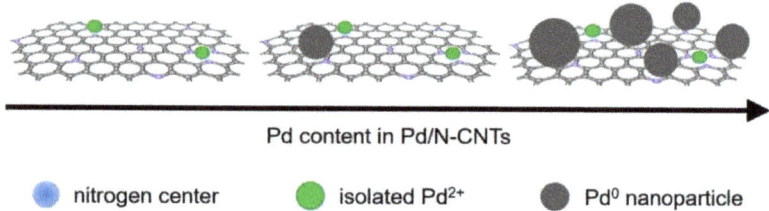

Figure 10. Schematic overview of the interaction of Pd with N-CNTs.

As shown by the catalytic experiments, isolated palladium ions are more active than metallic palladium nanoparticles, and the maximum TOF value is observed on the 0.2% Pd/N-CNT catalyst. A strong interaction of palladium with the pyridinic centers of N-CNTs is responsible for high stability of the catalyst in the course of the reaction as compared to the N-free catalyst (Figure 8c). Presumably, activity of the catalysts with high palladium content is determined by the ratio of two palladium species and by the size of metallic nanoparticles. In comparison with N-free catalysts, selectivity on all the N-doped catalysts increased by 5%, which corresponds to nearly a four-fold decrease in CO concentration. Additionally, the discovered stability of the catalyst 2% Pd/N-CNT compared to 2% Pd/CNT confirms the stabilization of both palladium species (isolated ions and metal nanoparticles) by nitrogen centers [49].

5. Conclusions

Nitrogen centers of N-CNTs were shown to be efficient in the stabilization of different palladium species in the form of isolated ions and highly dispersed metal nanoparticles that are active towards formic acid decomposition for hydrogen production. This allows a conclusion that various nitrogen-containing carbon nanomaterials are promising for use as the supports of metallic catalysts for this reaction. Taking into account an increased activity of isolated palladium ions, it is more preferable to use N-CNMs with predominant pyridinic nitrogen centers in the external graphene layers.

Author Contributions: Conceptualization, O.Y.P.; data curation, A.N.S. and O.Y.P.; synthesis, A.N.S.; investigation, A.N.S. and O.Y.P.; methodology, O.Y.P.; writing—review and editing, O.Y.P. All authors have read and agreed to the published version of the manuscript.

Funding: This work was supported by the Russian Science Foundation (Grant No. 17-73-30032).

Institutional Review Board Statement: Not applicable.

Informed Consent Statement: Not applicable.

Data Availability Statement: Data is contained within the article.

Acknowledgments: The authors thank D. Svintsitskiy, A. Lisitsyn, O. Stonkus, V. Sobolev and A. Chuvilin for the performance of some physicochemical studies. The authors also thank D. Bulushev for discussion.

Conflicts of Interest: The authors declare no conflict of interest.

References

1. Grasemann, M.; Laurenczy, G. Formic acid as a hydrogen source—Recent developments and future trends. *Energy Environ. Sci.* **2012**, *5*, 8171–8181. [CrossRef]
2. Bulushev, D.A.; Ross, J.R.H. Towards Sustainable Production of Formic Acid. *ChemSusChem* **2018**, *11*, 821–836. [CrossRef] [PubMed]
3. Tang, Y.; Roberts, C.A.; Perkins, R.T.; Wachs, I.E. Revisiting formic acid decomposition on metallic powder catalysts: Exploding the HCOOH decomposition volcano curve. *Surf. Sci.* **2016**, *650*, 103–110. [CrossRef]
4. Tang, C.; Surkus, A.; Chen, F.; Pohl, M.; Agostini, G.; Schneider, M.; Junge, H.; Beller, M. A stable nanocobalt catalyst with highly dispersed CoNx active sites for the selective dehydrogenation of formic acid. *Angew. Chem.* **2017**, *56*, 16616–16620. [CrossRef] [PubMed]
5. He, L.; Weniger, F.; Neumann, H.; Beller, M. Synthesis, Characterization, and Application of Metal Nanoparticles Supported on Nitrogen-Doped Carbon: Catalysis beyond Electrochemistry. *Angew. Chem. Int. Ed.* **2016**, *55*, 12582–12594. [CrossRef]
6. Shi, W.; Zhang, B.; Lin, Y.; Wang, Q.; Zhang, Q.; Su, D.S. Enhanced Chemoselective Hydrogenation through Tuning the Interaction between Pt Nanoparticles and Carbon Supports: Insights from Identical Location Transmission Electron Microscopy and X-ray Photoelectron Spectroscopy. *ACS Catal.* **2016**, *6*, 7844–7854. [CrossRef]
7. Ning, X.; Yu, H.; Peng, F.; Wang, H. Pt nanoparticles interacting with graphitic nitrogen of N-doped carbon nanotubes: Effect of electronic properties on activity for aerobic oxidation of glycerol and electro-oxidation of CO. *J. Catal.* **2015**, *325*, 136–144. [CrossRef]
8. Ombaka, L.M.; Ndungu, P.G.; Nyamori, V.O. Pyrrolic nitrogen-doped carbon nanotubes: Physicochemical properties, interactions with Pd and their role in the selective hydrogenation of nitrobenzophenone. *RSC Adv.* **2015**, *5*, 109–122. [CrossRef]
9. Perini, L.; Durante, C.; Favaro, M.; Perazzolo, V.; Agnoli, S.; Schneider, O.; Granozzi, G.; Gennaro, A. Metal–Support Interaction in Platinum and Palladium Nanoparticles Loaded on Nitrogen-Doped Mesoporous Carbon for Oxygen Reduction Reaction. *ACS Appl. Mater. Interfaces* **2015**, *7*, 1170–1179. [CrossRef]
10. Liu, W.; Chen, Y.; Qi, H.; Zhang, L.; Yan, W.; Liu, X.; Yang, X.; Miao, S.; Wang, W.; Liu, C.; et al. A Durable Nickel Single-Atom Catalyst for Hydrogenation Reactions and Cellulose Valorization under Harsh Conditions. *Angew. Chem.* **2018**, *57*, 7071–7075. [CrossRef]
11. Huang, X.; Xia, Y.; Cao, Y.; Zheng, X.; Pan, H.; Zhu, J.; Ma, C.; Wang, H.; Li, J.; You, R.; et al. Enhancing both selectivity and coking-resistance of a single-atom Pd1/C3N4 catalyst for acetylene hydrogenation. *Nano Res.* **2017**, *10*, 1302–1312. [CrossRef]
12. Vilé, G.; Albani, D.; Nachtegaal, M.; Chen, Z.; Dontsova, D.; Antonietti, M.; López, N.; Pérez-Ramírez, J. A Stable Single-Site Palladium Catalyst for Hydrogenations. *Angew. Chem.* **2015**, *54*, 11265–11269. [CrossRef] [PubMed]

13. Bulushev, D.A.; Zacharska, M.; Lisitsyn, A.S.; Podyacheva, O.Y.; Hage, F.S.; Ramasse, Q.M.; Bangert, U.; Bulusheva, L.G. Single Atoms of Pt-Group Metals Stabilized by N-Doped Carbon Nanofibers for Efficient Hydrogen Production from Formic Acid. *ACS Catal.* **2016**, *6*, 3442–3451. [CrossRef]
14. Golub, F.S.; Beloshapkin, S.; Gusel'Nikov, A.V.; Bolotov, V.A.; Parmon, V.N.; Bulushev, D.A. Boosting Hydrogen Production from Formic Acid over Pd Catalysts by Deposition of N-Containing Precursors on the Carbon Support. *Energies* **2019**, *12*, 3885. [CrossRef]
15. Nishchakova, A.D.; Bulushev, D.A.; Stonkus, O.A.; Asanov, I.P.; Ishchenko, A.V.; Okotrub, A.V.; Bulusheva, L.G. Effects of the Carbon Support Doping with Nitrogen for the Hydrogen Production from Formic Acid over Ni Catalysts. *Energies* **2019**, *12*, 4111. [CrossRef]
16. Podyacheva, O.; Lisitsyn, A.; Kibis, L.; Boronin, A.; Stonkus, O.; Zaikovskii, V.; Suboch, A.; Sobolev, V.; Parmon, V. Nitrogen Doped Carbon Nanotubes and Nanofibers for Green Hydrogen Production: Similarities in the Nature of Nitrogen Species, Metal–Nitrogen Interaction, and Catalytic Properties. *Energies* **2019**, *12*, 3976. [CrossRef]
17. Navlani-García, M.; Mori, K.; Salinas-Torres, D.; Kuwahara, Y.; Yamashita, H. New Approaches Toward the Hydrogen Production From Formic Acid Dehydrogenation Over Pd-Based Heterogeneous Catalysts. *Front. Mater.* **2019**, *6*, 44. [CrossRef]
18. Bulushev, D.A.; Bulusheva, L.G. Catalysts with single metal atoms for the hydrogen production from formic acid. *Catal. Rev.* **2021**, 1–40. [CrossRef]
19. Wei, S.; Li, A.; Liu, J.-C.; Li, Z.; Chen, W.; Gong, Y.; Zhang, Q.; Cheong, W.-C.; Wang, Y.; Zheng, L.; et al. Direct observation of noble metal nanoparticles transforming to thermally stable single atoms. *Nat. Nanotechnol.* **2018**, *13*, 856–861. [CrossRef]
20. Podyacheva, O.Y.; Cherepanova, S.V.; Romanenko, A.I.; Kibis, L.S.; Svintsitskiy, D.A.; Boronin, A.I.; Stonkus, O.A.; Suboch, A.N.; Puzynin, A.V.; Ismagilov, Z.R. Nitrogen doped carbon nanotubes and nanofibers: Composition, structure, electrical conductivity and capacity properties. *Carbon* **2017**, *122*, 475–483. [CrossRef]
21. Svintsitskiy, D.A.; Kibis, L.S.; Smirnov, D.A.; Suboch, A.N.; Stonkus, O.A.; Podyacheva, O.Y.; Boronin, A.I.; Ismagilov, Z.R. Spectroscopic study of nitrogen distribution in N-doped carbon nanotubes and nanofibers synthesized by catalytic ethylene-ammonia decomposition. *Appl. Surf. Sci.* **2018**, *435*, 1273–1284. [CrossRef]
22. Podyacheva, O.Y.; Bulushev, D.A.; Suboch, A.N.; Svintsitskiy, D.A.; Lisitsyn, A.S.; Modin, E.; Chuvilin, A.; Gerasimov, E.Y.; Sobolev, V.I.; Parmon, V.N. Highly Stable Single-Atom Catalyst with Ionic Pd Active Sites Supported on N-Doped Carbon Nanotubes for Formic Acid Decomposition. *ChemSusChem* **2018**, *11*, 3724–3727. [CrossRef] [PubMed]
23. Terrones, M.; Ajayan, P.; Banhart, F.; Blase, X.; Carroll, D.; Charlier, J.-C.; Czerw, R.; Foley, B.; Grobert, N.; Kamalakaran, R.; et al. N-doping and coalescence of carbon nanotubes: Synthesis and electronic properties. *Appl. Phys. A* **2002**, *74*, 355–361. [CrossRef]
24. Chizari, K.; Janowska, I.; Houllé, M.; Florea, I.; Ersen, O.; Romero, T.; Bernhardt, P.; LeDoux, M.J.; Pham-Huu, C. Tuning of nitrogen-doped carbon nanotubes as catalyst support for liquid-phase reaction. *Appl. Catal. A Gen.* **2010**, *380*, 72–80. [CrossRef]
25. Van Dommele, S.; Romero-Izquirdo, A.; Brydson, R.; De Jong, K.; Bitter, J. Tuning nitrogen functionalities in catalytically grown nitrogen-containing carbon nanotubes. *Carbon* **2008**, *46*, 138–148. [CrossRef]
26. Bulusheva, L.G.; Okotrub, A.V.; Fedoseeva, Y.V.; Kurenya, A.G.; Asanov, I.P.; Vilkov, O.Y.; Koós, A.A.; Grobert, N. Controlling pyridinic, pyrrolic, graphitic, and molecular nitrogen in multi-wall carbon nanotubes using precursors with different N/C ratios in aerosol assisted chemical vapor deposition. *Phys. Chem. Chem. Phys.* **2015**, *17*, 23741–23747. [CrossRef] [PubMed]
27. Tao, X.; Zhang, X.; Sun, F.; Cheng, J.; Liu, F.; Luo, Z. Large-scale CVD synthesis of nitrogen-doped multi-walled carbon nanotubes with controllable nitrogen content on a $Co_xMg_{1-x}MoO_4$ catalyst. *Diam. Relat. Mater.* **2007**, *16*, 425–430. [CrossRef]
28. Kiciński, W.; Dyjak, S. Transition metal impurities in carbon-based materials: Pitfalls, artifacts and deleterious effects. *Carbon* **2020**, *168*, 748–845. [CrossRef]
29. Gulino, G.; Vieira, R.; Amadou, J.; Nguyen, P.; LeDoux, M.J.; Galvagno, S.; Centi, G.; Pham-Huu, C. C_2H_6 as an active carbon source for a large scale synthesis of carbon nanotubes by chemical vapour deposition. *Appl. Catal. A Gen.* **2005**, *279*, 89–97. [CrossRef]
30. Guellati, O.; Begin, D.; Antoni, F.; Moldovan, S.; Guerioune, M.; Pham-Huu, C.; Janowska, I. CNTs' array growth using the floating catalyst-CVD method over different substrates and varying hydrogen supply. *Mater. Sci. Eng. B* **2018**, *231*, 11–17. [CrossRef]
31. Terrones, M.; Benito, A.; Manteca-Diego, C.; Hsu, W.; Osman, O.; Hare, J.; Reid, D.; Cheetham, A.; Prassides, K.; Kroto, H.; et al. Pyrolytically grown $B_xC_yN_z$ nanomaterials: Nanofibres and nanotubes. *Chem. Phys. Lett.* **1996**, *257*, 576–582. [CrossRef]
32. Han, W.-Q.; Kohler-Redlich, P.; Seeger, T.; Ernst, F.; Ruhle, M.; Grobert, N.; Hsu, W.-K.; Chang, B.-H.; Zhu, Y.-Q.; Kroto, H.W.; et al. Aligned CN_x nanotubes by pyrolysis of ferrocene/C_{60} under NH_3 atmosphere. *Appl. Phys. Lett.* **2000**, *77*, 1807–1809. [CrossRef]
33. Lobiak, E.V.; Kuznetsova, V.R.; Makarova, A.A.; Okotrub, A.V.; Bulusheva, L.G. Structure, functional composition and electrochemical properties of nitrogen-doped multi-walled carbon nanotubes synthesized using Co–Mo, Ni–Mo and Fe–Mo catalysts. *Mater. Chem. Phys.* **2020**, *255*, 123563. [CrossRef]
34. Choi, H.C.; Park, J.; Kim, B. Distribution and Structure of N Atoms in Multiwalled Carbon Nanotubes Using Variable-Energy X-Ray Photoelectron Spectroscopy. *J. Phys. Chem. B* **2005**, *109*, 4333–4340. [CrossRef]
35. Dresselhaus, M.; Dresselhaus, G.; Saito, R.; Jorio, A. Raman spectroscopy of carbon nanotubes. *Phys. Rep.* **2005**, *409*, 47–99. [CrossRef]

36. Dresselhaus, M.S.; Jorio, A.; Filho, A.G.S.; Saito, R. Defect characterization in graphene and carbon nanotubes using Raman spectroscopy. *Philos. Trans. R. Soc. A Math. Phys. Eng. Sci.* **2010**, *368*, 5355–5377. [CrossRef] [PubMed]
37. Podila, R.; Chacón-Torres, J.; Spear, J.T.; Pichler, T.; Ayala, P.; Rao, A.M. Spectroscopic investigation of nitrogen doped graphene. *Appl. Phys. Lett.* **2012**, *101*, 123108. [CrossRef]
38. Golubtsov, G.V.; Kazakova, M.A.; Selyutin, A.G.; Ishchenko, A.V.; Kuznetsov, V.L. Mono-, Bi-, and Trimetallic Catalysts for the Synthesis of Multiwalled Carbon Nanotubes Based on Iron Subgroup Metals. *J. Struct. Chem.* **2020**, *61*, 640–651. [CrossRef]
39. Fenelonov, V.; Derevyankin, A.; Okkel, L.; Avdeeva, L.; Zaikovskii, V.; Moroz, E.; Salanov, A.; Rudina, N.; Likholobov, V.; Shaikhutdinov, S. Structure and texture of filamentous carbons produced by methane decomposition on NI and NI-CU catalysts. *Carbon* **1997**, *35*, 1129–1140. [CrossRef]
40. Kumar, K.V.; Preuss, K.; Guo, Z.X.; Titirici, M.M. Understanding the Hydrophilicity and Water Adsorption Behavior of Nanoporous Nitrogen-Doped Carbons. *J. Phys. Chem. C* **2016**, *120*, 18167–18179. [CrossRef]
41. Chernyak, S.; Burtsev, A.; Maksimov, S.; Kupreenko, S.; Maslakov, K.; Savilov, S. Structural evolution, stability, deactivation and regeneration of Fischer-Tropsch cobalt-based catalysts supported on carbon nanotubes. *Appl. Catal. A Gen.* **2020**, *603*, 117741. [CrossRef]
42. Hao, G.-P.; Sahraie, N.R.; Zhang, Q.; Krause, S.; Oschatz, M.; Bachmatiuk, A.; Strasser, P.; Kaskel, S. Hydrophilic non-precious metal nitrogen-doped carbon electrocatalysts for enhanced efficiency in oxygen reduction reaction. *Chem. Commun.* **2015**, *51*, 17285–17288. [CrossRef]
43. Bueres, R.F.; Asedegbega-Nieto, E.; Díaz, E.; Ordóñez, S.; Diez, F.V. Performance of carbon nanofibres, high surface area graphites, and activated carbons as supports of Pd-based hydrodechlorination catalysts. *Catal. Today* **2010**, *150*, 16–21. [CrossRef]
44. Lesiak, B.; Mazurkiewicz, M.; Malolepszy, A.; Stobinski, L.; Mierzwa, B.; Mikolajczuk-Zychora, A.; Juchniewicz, K.; Borodzinski, A.; Zemek, J.; Jiricek, P. Effect of the Pd/MWCNTs anode catalysts preparation methods on their morphology and activity in a direct formic acid fuel cell. *Appl. Surf. Sci.* **2016**, *387*, 929–937. [CrossRef]
45. Fleisch, T.; Zajac, G.; Schreiner, J.; Mains, G. An XPS study of the UV photoreduction of transition and noble metal oxides. *Appl. Surf. Sci.* **1986**, *26*, 488–497. [CrossRef]
46. Arrigo, R.; Schuster, M.E.; Xie, Z.; Yi, Y.; Wowsnick, G.; Sun, L.L.; Hermann, K.E.; Friedrich, M.; Kast, P.; Hävecker, M.; et al. Nature of the N–Pd Interaction in Nitrogen-Doped Carbon Nanotube Catalysts. *ACS Catal.* **2015**, *5*, 2740–2753. [CrossRef]
47. He, Z.; Dong, B.; Wang, W.; Yang, G.; Cao, Y.; Wang, H.; Yang, Y.; Wang, Q.; Peng, F.; Yu, H. Elucidating Interaction between Palladium and N-Doped Carbon Nanotubes: Effect of Electronic Property on Activity for Nitrobenzene Hydrogenation. *ACS Catal.* **2019**, *9*, 2893–2901. [CrossRef]
48. Arrigo, R.; Wrabetz, S.; Schuster, M.E.; Wang, D.; Villa, A.; Rosenthal, D.; Girsgdies, F.; Weinberg, G.; Prati, L.; Schlögl, R.; et al. Tailoring the morphology of Pd nanoparticles on CNTs by nitrogen and oxygen functionalization. *Phys. Chem. Chem. Phys.* **2012**, *14*, 10523. [CrossRef] [PubMed]
49. Cao, Y.; Mao, S.; Li, M.; Chen, Y.; Wang, Y. Metal/Porous Carbon Composites for Heterogeneous Catalysis: Old Catalysts with Improved Performance Promoted by N-Doping. *ACS Catal.* **2017**, *7*, 8090–8112. [CrossRef]

Article

Formic Acid as a Hydrogen Donor for Catalytic Transformations of Tar

Vladimir V. Chesnokov *, Pavel P. Dik and Aleksandra S. Chichkan

Boreskov Institute of Catalysis, Siberian Branch, Russian Academy of Sciences, 630090 Novosibirsk, Russia; dik@catalysis.ru (P.P.D.); alexcsh@yandex.ru (A.S.C.)
* Correspondence: chesn@catalysis.ru; Tel.: +7-383-326-9724

Received: 3 August 2020; Accepted: 26 August 2020; Published: 1 September 2020

Abstract: Specific features of the catalytic tar cracking in the presence of formic acid, BEA zeolite and 8% Ni-2.5% Mo/Sibunit catalyst were studied at 350 °C and 1.0 MPa pressure. The obtained results evidenced that formic acid can be used as a hydrogen donor during catalytic reactions. The formic acid addition made it possible to perform efficient hydrocracking of heavy feed such as tar. It was found that both the tar conversion and selectivity to light (gasoline-diesel) fractions grew in the sequence: tar < (tar - formic acid) < (tar - formic acid - BEA zeolite) < (tar - formic acid - BEA zeolite - 8% Ni-2.5% Mo/Sibunit catalyst). Furthermore, significantly lower concentrations of impurities containing sulfur and nitrogen were observed for the (tar - formic acid - BEA zeolite - 8% Ni-2.5% Mo/Sibunit catalyst) system. For example, the sulfur and nitrogen concentrations in the tar precursor were 1.50% and 0.86%, respectively. Meanwhile, their concentrations in the liquid products after the catalytic cracking were 0.73% and 0.18%, respectively.

Keywords: hydrocracking; tar; formic acid; nickel; zeolite; hydrogen donor; catalyst

1. Introduction

Due to increasing worldwide demand for motor fuel, the need to utilize non-traditional heavy oil feedstock (HOF), including heavy oils, natural bitumen, heavy residual oil fractions (tar, black oil), bituminous sands and pyroshales is also rapidly growing. A number of methods for deep processing of heavy oil fractions, heavy oils and oil residues in the presence of catalysts have been reported. Traditional approaches to HOF processing can be divided into two main types [1–4]. The first type includes deasphalting, thermal processes: gasification and carbonization (delayed, flexicoking, etc.), viscosity breaking and catalytic cracking. The second type includes various hydrogenation processes. Therefore, there is great interest in hydrocracking of heavy residual oil fractions, particularly tar, aimed at the production of the main products of the gasoline and diesel fractions that are in great demand. The chemistry of hydrogenation processes is based on the efficient use of hydrogen. However, reactors used for hydrocracking processes operate under high pressure and require complex instrumentation. Thus, researchers are currently looking for new sources of hydrogen.

One of the actively developing areas combining the advantages of the thermal and hydrogenation refinement processes is catalytic vapor cracking of HOF [5–7]. In this process water acts as a hydrogen donor. Its participation makes it possible to increase the yield of the light fraction (boiling point below 360 °C) and decrease the coke yield as well as the contents of sulfur and heteroatoms in the liquid products. The use of catalysts based on Ni [8,9], Fe [10,11], Mo [8,12], Co [13] and Zr [14,15] in the catalytic vapor cracking favors stronger HOF interaction with water, including partial oxidation, low-temperature partial steam reforming and catalytic cracking processes, thus substantially improving the overall efficiency of the process.

A different approach was demonstrated in other publications. It was suggested to add methane to CO_2 for cracking of resins obtained by coal pyrolysis. Due to its high H/C ratio, methane is considered to be a hydrogen substitute. Hydrogen is formed by methane reforming with carbon dioxide [16,17].

However, these approaches are complicated for practical implementation because two complex processes have to be optimized: hydrogen production and the hydrocracking itself. The use of formic acid as a hydrogen donor significantly simplifies the process. Formic acid is a high-quality organic hydrogen source, as it has relatively high hydrogen content (4.4 wt.%), low flammability and low toxicity. It is important to note that formic acid can be prepared from the biomass or from CO_2 [18–20].

Delayed carbonization is the most popular scheme for processing of tar and other heavy residues. In the current study, modification of the tar carbonization was attempted. The formic acid addition was suggested as a means to regulate the yield and properties of the formed carbonization products. The goal of the current study was to investigate the effect of the formic acid addition and different catalysts on the yield and properties of liquid tar hydrocracking products.

2. Materials and Methods

Tar from the Omsk Oil Refinery (Omsk, Russia) was used as a feedstock. The elemental and fraction compositions of the used tar are reported in Table 1.

Table 1. Characteristics of the studied tar.

Elemental Composition	Concentration (wt.%)
C	86.8
H	11.7
N	0.86
S	1.50
Fraction Composition, wt.%	
Gasoline fractions (<200 °C)	0
Diesel fractions (200–350 °C)	0.1
Vacuum gas oil (350–500 °C)	6.7
Tar (500–700 °C)	54.9
Tar (>700 °C)	38.2

Hydrocracking catalysts are bifunctional. Their hydrogenation–dehydrogenation function and activity in all reactions related to hydrorefining is usually associated with MoS_2 or WS_2 promoted with nickel sulfide. Amorphous or crystalline aluminosilicates, Y [21] or BEA [22–24] zeolites are typically responsible for the cracking function. Based on this information, a nickel–molybdenum catalyst and a BEA zeolite were used in this study for the tar hydrocracking.

The use of carbon supports for tar carbonization is desirable because they do not increase the ash content of the resulting oil coke. The artificial mesoporous carbon support "Sibunit" [25,26], which has a surface of 480 m^2/g and a graphite-like structure, was used for the catalyst synthesis. A TEM image of the Sibunit support is shown in Figure 1. The support particles have an egg-shell shape. Sibunit is a graphite-like carbon material with an interlayer distance of d_{002} = 3.52 Å and an average crystallite size in this direction of about 25 Å.

The 8% Ni-2.5% Mo/Sibunit catalyst was prepared by impregnation. Precalculated amounts of $Ni(NO_3)_2 \cdot 6H_2O$ and $(NH_4)_2MoO_4$ (Reakhim, 98% purity) were dissolved in 10 mL of distilled water and added to a beaker containing 10 g of Sibunit. Then, the sample was dried on a hot plate with a magnetic stirrer and calcined in a muffle furnace in air at 250 °C for 30 min. The obtained NiO-MoO/Sibunit sample was reduced in the hydrogen flow in a flow reactor at 400 °C for 1 h. After reduction the catalyst composition was 8% Ni-2.5% Mo/Sibunit. The XRD pattern of 8% Ni-2.5% Mo/Sibunit reduced in hydrogen is shown in Figure 2.

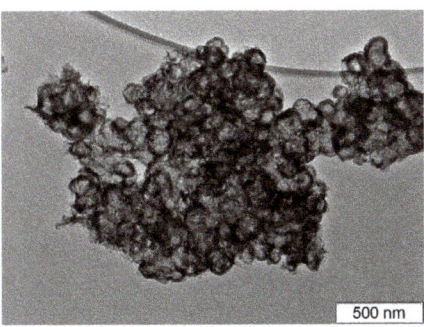

Figure 1. TEM image of the Sibunit support.

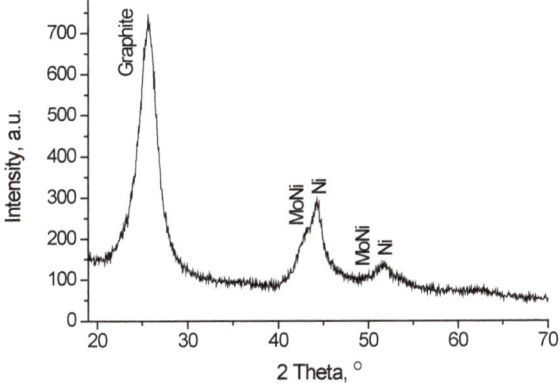

Figure 2. XRD pattern of 8% Ni-2.5% Mo/Sibunit.

The peaks observed in the XRD pattern of 8% Ni-2.5% Mo/Sibunit can be attributed to nickel metal or a nickel–molybdenum phase (Figure 2). The peak at $2\theta \sim 44°$ is formed by overlapping of different peaks and has a shoulder. This shoulder ($2\theta \sim 42$–$43°$) can be attributed either to graphite or to a nickel–molybdenum phase. The position of the peak close to the shoulder ($2\theta \sim 44.3°$) is typical for the nickel metal phase. The diffraction characteristics of the active component of the Ni-Mo/Sibunit catalyst are reported in Table 2.

Table 2. Diffraction characteristics of the active component of supported Ni-Mo/Sibunit.

Catalyst	Phase	Description	Average Cystallite Size D, Å	Elementary Cell Parameter, Å
Ni-Mo/Sibunit	Ni	Ni PDF 04-0850,	75	Fm3m, $a = 3.523$
	Ni-Mo	$Mo_{1.08}Ni_{2.92}$	60	Fm3m, $a = 3.637$

BEA zeolite ($SiO_2/Al_2O_3 = 27$) produced by AZKiOS (Angarsk, Russia) was used. Its characteristics are reported in Tables 3 and 4.

Table 3. Characteristics of the BEA zeolite.

No.	Parameter	Result
1	Na_2O content normalized to the weight of the sample calcined at 650 °C, wt.%	0.01
2	Al_2O_3 content normalized to the weight of the sample calcined at 650 °C, wt.%	5.6
3	SiO_2 content normalized to the weight of the sample calcined at 650 °C, wt.%	90
4	Crystallinity, %	86
5	SiO_2/Al_2O_3 ratio	27

Table 4. Textural characteristics of the BEA zeolite.

Zeolite	V_{micro}, cm³/g	V_{meso}, cm³/g	V_{total}, cm³/g	S_{BET}, m²/g	Average Size of the Crystals, nm
BEA	0.22	0.15	0.37	660	160

Catalytic reactions of tar were performed in an autoclave at 350–500 °C and 1 MPa pressure. A scheme of the autoclave unit for catalytic reactions of tar is shown in Figure 3. A tar sample (~16 g) or a tar sample (~16 g) with a catalyst (0.4 g) was loaded in the sample holder basket. In another series of experiments, a tar sample (~16 g) with formic acid (0.4 g) or a tar sample (~16 g) with formic acid (9 g) and a catalyst (0.4 g) was loaded in the sample holder basket. Then, the autoclave was heated to the desired temperature and maintained at it for 2 h.

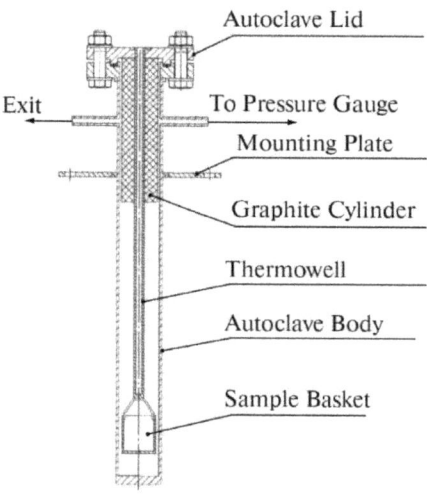

Figure 3. Scheme of the autoclave unit for catalytic reactions of tar.

The gas phase components were analyzed using a Kristall-2000M gas chromatograph (Chromatec, Yoshkar-Ola, Russia). A thermal conductivity detector was used for analysis of H_2S, COS, SO_2, H_2, O_2, CO_2 and CH_4 after their separation on a packed column (2 mm × 1.5 m) filled with SKT activated carbon "AQUACHEM" Kazan. A flame ionization detector was used for analysis of the gas phase organic components after separation on a packed column (2 mm × 3 m) filled with Hayesep Q + 0.9% PTMSP (Hayes Separations, Inc. of Bandera, TX, USA).

The concentrations of sulfur and nitrogen and the H:C ratio in the initial tar and products of its processing were determined using a VARIO EL CUBE CHNS-O-analyzer (Elementar Analysensysteme GmbH, Langenselbold, Germany). The fraction compositions of the tar and products formed from it were determined by imitation distillation according to ASTM D7169 using an Agilent 7890B gas chromatograph (Santa Clara, CA, USA).

The elemental compositions of solid phases were determined using the X-ray fluorescence spectrometer ARL Perform'X (Bruker, Germany) with a Rh anode for fluorescence excitation. Powder samples were ground in an agate mortar to a homogeneous finely dispersed state. Powder and liquid (viscous) samples were placed into a special holder covered with a Spectrolenesix polypropylene film that is transparent to the X-ray irradiation. Then, this holder was placed into the spectrometer chamber filled with helium. Automatic spectrum smoothing, background subtraction and calculation of the concentrations were performed using the UniQuant software package (Germany).

The phase composition of the samples was studied by the X-ray diffraction analysis. Diffraction patterns were recorded with a Thermo ARL X'TRA diffractometer (Thermo Fisher Scientific, Basel, Switzerland) using CuKα radiation with a wavelength of 1.54184 Å. The 2θ scan range was 5–75°, the scanning step was 0.05° and the accumulation time at each point was 5 s. Diffraction (Powder Diffraction Files) (PDF) and structural (Inorganic Crystal Structure Database) (ICSD) databases were used to identify the phase compositions. The average crystallite sizes were determined using the Selyakov–Scherrer formula from the integral widths of the diffraction peaks.

High-resolution transmission electron microscopy (HRTEM) images were obtained using a JEOL JEM-2010 (Tokyo, Japan) electron microscope with a lattice resolution of 0.14 nm.

The experiments on the formic acid (FA) decomposition were performed in a flow installation using a quartz reactor (inner diameter 6 mm). The catalyst loading was 20 mg. The catalyst was uniformly mixed with 0.5 cm^3 of quartz. The feed consisted of 5 vol.% FA in helium. Its flow rate was 20 cm^3/min. The experiments were performed in a temperature-programmed mode with a temperature increase rate of 2 °C/min. The reaction progress was followed by monitoring the release of CO and CO_2.

3. Results and Discussion

3.1. Thermal Transformations of Tar

The tar carbonization was studied in the temperature range of 350–500 °C. At 350 °C, the tar carbonization yielded gaseous and liquid products as well as oil coke. The amount of liquid product was about 15 wt.%. At 450 °C only gaseous and solid carbonization products were formed. The temperature increase favors conversion of liquid products into gases. The composition of gaseous products obtained at 350–500 °C is reported in Table 5. Methane, ethane and propane were the main gaseous carbonization products. The methane concentration in the products grew with carbonization time.

Table 5. The effect of tar carbonization temperature on the composition of gaseous products (reaction time: 1 h).

Hydrocarbons	Temperature, °C			
	350	400	450	500
	Concentration, vol.%			
H_2	7	8.0	15.4	18.5
CH_4	36	41.8	48.1	65
C_2H_6	24.5	21.2	15.4	12
C_2H_4	4	3.1	2.8	1.5
C_3H_8	12	11.8	6.6	0.6
C_3H_6	0.2	0.2	0.1	0
iso-C_4H_{10}	4.5	4.0	4.0	0.5
n-C_4H_{10}	2.0	1.0	0.5	0.1
1-butene	4.0	3.9	2.6	0.4
C_{5+}	4.8	4.2	3.8	0.8
C_{6+}	1.0	0.8	0.7	0.6

Significant amounts of H₂S and COS were observed in the reaction products. The contents of the sulfur-containing compounds in gaseous products formed during the tar carbonization at 450 °C are reported in Table 6. Hydrocarbon gases, as well as sulfur-containing H₂S and COS products, were formed during the process. Gaseous products were periodically removed from the autoclave, and therefore a release of weakly bound sulfur from the tar decreased as the coking process proceeded. This explains the decreasing concentration of sulfur-containing products in the gas as the reaction time increased.

Table 6. Concentrations of sulfur-containing compounds in gaseous products of tar carbonization at 450 °C as a function of reaction time.

Conditions of Carbonization	Reaction Time, h	1	2	3	5
Tar carbonization without a catalyst	H₂S content, wt.%	1.44	1.58	0.82	0.61
	COS content, wt.%	0.22	2.15	2.15	0.75

3.2. Catalytic Decomposition of Formic Acid

Hydrogen can be produced from formic acid by a catalytic or non-catalytic reaction. The formic acid decomposition in the gas phase can follow two pathways: dehydrogenation with the formation of H₂ and CO₂ (Equation (1)), and dehydration with the formation of CO and H₂O (Equation (2)).

$$\text{HCOOH}_{(gas)} \rightarrow H_2 + CO_2, \Delta_r H^\circ_{298} = -14.7 \frac{kj}{mol} \quad (1)$$

$$\text{HCOOH}_{(gas)} \rightarrow H_2O_{(gas)} + CO, \Delta_r H^\circ_{298} = 26.5 \frac{kj}{mol} \quad (2)$$

Formic acid decomposition can be catalyzed by a number of metals. Palladium catalysts are the most efficient for the hydrogen production from formic acid according to Equation (1) [27–31]. Nickel catalysts have lower activity [32]. However, from the economic viewpoint nickel catalysts deserve more attention than noble metal catalysts. The catalyst supports play an important role in the catalytic activity, especially when a highly dispersed active metal is used. Porous carbon materials are suitable candidates, since they have all the properties required for such support.

Experiments in a flow installation with a quartz reactor were carried out to determine the pathways of the formic acid decomposition over the 8% Ni-2.5% Mo/Sibunit catalyst. The 8% Ni-2.5% Mo/Sibunit catalyst was pre-treated in a flow of 5 vol.% FA in helium at 300 °C to remove the oxide layers formed by contact of metal particles with air. The FA conversion was calculated as a ratio of the sum of the CO and CO₂ concentrations to the initial FA concentration. The selectivity to CO₂ (H₂) was determined as a ratio of the CO₂ concentration to the sum of the CO and CO₂ concentrations. The results of the formic acid catalytic decomposition over the 8% Ni-2.5% Mo/Sibunit catalyst are presented in Figure 4.

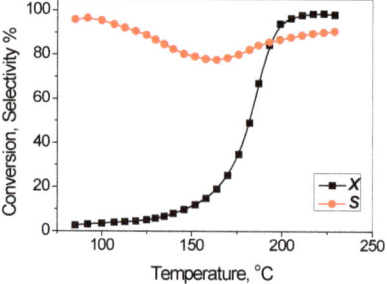

Figure 4. FA decomposition on 8% Ni-2.5% Mo/Sibunit catalyst after pretreatment in FA at 300 °C for 1 h.

The formic acid decomposition on the 8% Ni-2.5% Mo/Sibunit catalyst preferentially follows the pathway of Equation (1) with the formation of CO_2 and H_2 (Figure 4). Similar experiments on the formic acid decomposition were performed in an empty autoclave without a catalyst. About 3 mL of formic acid was placed into a 300 mL autoclave. The autoclave was gradually heated to 250 °C and maintained at this temperature for 30 min. Then, the temperature was increased in several steps to 350 °C. It was found that the pressure growth in the autoclave began at 120 °C. At 250 °C the pressure was equal to 4 atm. The gas phase composition was studied by gas chromatography. The dependence of the gas phase composition and pressure on temperature is reported in Table 7. Even in the empty autoclave the formic acid decomposition followed the pathway of Equation (1). This is most likely caused by the catalytic effect of the steel autoclave walls on the formic acid decomposition. A CO methanation reaction also occurred to some extent. The obtained results evidenced that formic acid can be used as a source of hydrogen during catalytic reactions of tar. The change in pressure at 250 and 300 °C occurs because the system needs time to reach a steady state.

Table 7. Changes of the gas phase composition and pressure in the autoclave during the formic acid decomposition as a function of temperature.

Compound	Temperature, °C		
	250	300	350
	Concentration, vol.%		
Methane	0.6	0.8	0.9
Hydrogen	10–36	42–45	45–46
CO	12–13	7–6	6
Pressure, atm	4–8	8–13	14

It should be noted that products of formic acid decomposition might be in equilibrium with each other. The water–gas shift (WGS) reaction described in Equation (3) is an industrial process in which water in the form of steam is mixed with carbon monoxide to obtain hydrogen and carbon dioxide.

$$H_2O + CO = CO_2 + H_2 \tag{3}$$

The WGS reaction is reversible and exothermic ($\Delta H° = -41.2$ kJ/mol). The WGS reaction is thermodynamically favorable at temperatures of 300–400 °C [33]. Iron-based catalysts are typically used industrially. Therefore, the water–gas shift (WGS) reaction also favors the formation of hydrogen and CO_2.

3.3. Effect of Formic Acid on the Catalytic Transformations of Tar

The data on the tar, FA and catalyst loadings, and the selectivity to different fractions are summarized in Table 8. The reaction was performed at 350 °C for 2 h.

In addition to the total tar conversion, it is important to control the composition of liquid products. They were analyzed using the imitation distillation method. The fractions of gasoline (0–180 °C) and diesel (180–360 °C) fractions, vacuum gasoil (360–550 °C), vacuum residue (550–720 °C) and non-eluted residue were calculated according to ASTM D7169 on Agilent 7890B. The fraction compositions of the obtained liquid products are presented in Table 9 and Figure 5.

Table 8. The influence of formic acid and catalysts on the tar transformations at 350 °C.

Sample	Loading	Mass, g	The Products Composition, wt.%		
			Gas *	Liquid Hydrocarbons	Unconverted Tar **
1	Tar	16	8	15	77
2	Tar FA	16 9	11	43	46
3	Tar FA BEA zeolite	16 9 0.4	12	50	38
4	Tar FA Ni-Mo/Sibunit	16 9 0.4	17	48	35
5	Tar FA Ni-Mo/Sibunit BEA zeolite	16 9 0.2 0.2	17	55	28
			Without FA		
6	Tar Ni-Mo/Sibunit BEA zeolite	16 0.2 0.2	15	35	50

* The amount of formed gases without gaseous products of FA decomposition. ** Without weight of catalysts.

Table 9. Fraction distribution of liquid product formed by tar conversion in different systems.

		Sample					
		1	2	3	4	5	6
Fraction	Temperature, °C	Tar	Tar, FA	Tar, FA, BEA zeolite	Tar, FA, Ni-Mo/Sibunit	Tar, FA, Ni-Mo/Sibunit, BEA Zeolite	Tar, Ni-Mo/Sibunit, BEA Zeolite
		Concentration, wt.%					
Gasoline	0–180	10.8	12	16	12.7	19.2	7.7
Diesel	180–360	39.1	52.1	53.9	56.3	60.2	73.4
Vacuum gasoil	360–550	35.5	24.3	20.4	16.1	14.7	9.8
Vacuum residue	550–720	8.1	6.6	4.9	7.9	3.1	1.7
Non-eluted residue	>720	6.5	5	4.8	7.0	2.8	7.4

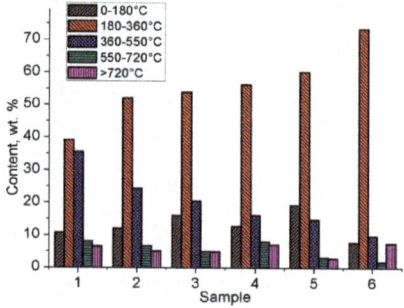

Figure 5. Fraction composition of liquid products.

The presented data demonstrate that the addition of formic acid and the catalysts resulted in redistribution of the liquid products between different fractions. The lowest amount of the gasoline (10.8%) and diesel (39.1%) fractions was observed in the liquid products formed by the thermal tar cracking. The addition of formic acid alone or together with the BEA zeolite to tar increased the contributions of these fractions among the liquid products.

Substitution of the BEA zeolite for the 8% Ni-2.5% Mo/Sibunit catalyst led to a decrease of the gasoline fraction concentration from 16% to 12.7%. This means that the 8% Ni-2.5% Mo/Sibunit catalyst has lower cracking ability than the BEA zeolite. Nickel metal is known to be a good catalyst for methanation and hydrogenolysis of C-C bonds [34–37]. These properties lead to its higher selectivity to gaseous products (Table 8). Meanwhile, simultaneous addition of the BEA zeolite and the 8% Ni-2.5% Mo/Sibunit catalyst (sample 380) led to the highest yields of both the gasoline and the diesel fractions.

Changes of the sulfur and nitrogen contents in the liquid reaction products were also studied. The results are presented in Table 10. Variations of the sulfur and nitrogen concentrations were symbatic, most likely due to similarities in the mechanisms of desulfurization and denitrogenation of liquid products during the catalytic transformations of tar.

Table 10. Changes of the sulfur and nitrogen concentrations in liquid products depending on the addition of formic acid and the catalysts.

Elements	Sample				
	1	2	3	4	5
	Tar	Tar, FA	Tar, FA, BEA Zeolite	Tar, FA, Ni-Mo/Sibunit	Tar, FA, Ni-Mo/Sibunit, BEA Zeolite
	Concentration, wt.%				
Sulfur	1.0	0.98	0.83	0.77	0.73
Nitrogen	0.39	0.33	0.22	0.2	0.18

The lowest sulfur concentration was observed for sample 5. This result was observed despite the fact that the original carbon support Sibunit contained 0.5 wt.% S. Ni-Mo catalysts are known to be among the best catalysts for purification of diesel fuel from sulfur. When the BEA zeolite is present together with the 8% Ni-2.5% Mo/Sibunit catalyst, it also contributes to desulfurization of the liquid products.

4. Conclusions

Thermal cracking of tar was studied at 350–500 °C and 1.0 MPa pressure. At 350 °C the tar carbonization resulted in the formation of gaseous and liquid products as well as oil coke. The fraction of liquid products was about 15 wt.%. Only gaseous and solid products were observed when the carbonization temperature was increased to 450 °C. Apparently, the temperature increase led to the conversion of liquid products to gases. The main gaseous tar carbonization products were methane, ethane and propane. H_2S and COS were observed in the reaction products. Due to this fact, the sulfur concentration on the oil coke decreased in comparison with the precursor tar from 1.5 to 1.28 wt.%.

Specific features of the catalytic tar cracking in the presence of formic acid, BEA zeolite and the 8% Ni-2.5% Mo/Sibunit catalyst were studied at 350 °C 1.0 MPa pressure. The experiments showed that the formic acid addition made it possible to perform efficient hydrocracking of heavy feeds such as tar. Both the tar conversion and selectivity to light (gasoline–diesel) fractions increased in the sequence: tar < (tar - formic acid) < (tar - formic acid - BEA zeolite) < (tar - formic acid - BEA zeolite - 8% Ni-2.5% Mo/Sibunit catalyst). Furthermore, significantly lower concentrations of impurities containing sulfur and nitrogen were observed for the (tar - formic acid - BEA zeolite - 8% Ni-2.5% Mo/Sibunit catalyst) system. For example, the sulfur and nitrogen concentrations in the tar precursor were 1.50% and 0.86%, respectively. Meanwhile, their concentrations in the liquid products after the catalytic cracking were 0.73% and 0.18%, respectively.

Author Contributions: Conceptualization, V.V.C.; methodology, V.V.C.; writing—original draft preparation, V.V.C.; investigation, P.P.D.; characterization, P.P.D.; measurements, P.P.D.; synthesis, A.S.C.; measurements, A.S.C.; data analysis, A.S.C. All authors have read and agreed to the published version of the manuscript.

Funding: This study was supported by the Russian Science Foundation, project 17–73–30032.

Conflicts of Interest: The authors declare no conflict of interest.

Nomenclature

HOF	heavy oil feedstock
HRTEM	high resolution transmission electron microscopy
PDF	Powder Diffraction Files
ICSD	Inorganic Crystal Structure Database
BET	Brunauer–Emmett–Teller
XRD	X-ray diffraction
WGS	water–gas shift reaction

References

1. Ancheyta, J.; Speight, J.G. *Hydroprocessing of Heavy Oils and Residua*; CRC Press: Boca Raton, FL, USA, 2007.
2. Surkov, V.G.; Pevneva, G.S.; Golovko, A.K. Structural and chemical transformations of asphaltenes and tar resins under conditions of mechanical influence. *Neftepererab. Neftekhim.* **2015**, *12*, 6–10.
3. Ruiz, J.C.S. *Applied Industrial Catalysis*; Archer Press: New York, NY, USA, 2017.
4. Gary, J.H.; Handwerk, G.E.; Kaiser, M.J. *Oil Refining: Technology and Economics*, 5th ed.; CRS Press: New York, NY, USA, 2007.
5. Zaikina, O.O.; Saiko, A.V.; Sosnin, G.A.; Yeletsky, P.M.; Gulyaeva, Y.K.; Klimov, O.V.; Noskov, A.S.; Yakovlev, V.A. Investigation of the properties of semisynthetic oils obtained in the presence of dispersed catalysts based on Mo and Fe-Co in the process of catalytic steam cracking of vacuum residue. *J. Sib. Fed. Univ. Chem.* **2019**, *12*, 512–521. [CrossRef]
6. Eletskii, P.M.; Mironenko, O.O.; Kukushkin, R.G.; Sosnin, G.A.; Yakovlev, V.A. Catalytic steam cracking of heavy oil feedstocks: A review. *Catal. Ind.* **2018**, *10*, 185–201. [CrossRef]
7. Eletskii, P.M.; Sosnin, G.A.; Zaikina, O.O.; Kukushkin, R.G.; Yakovlev, V. Heavy oil upgrading in the presence of water. *J. Sib. Fed. Univ. Chem.* **2018**, *10*, 545–572. [CrossRef]
8. Reina, T.R.; Yeletsky, P.; Bermúdez, J.M.; Arcelus-Arrillaga, P.; Yakovlev, V.A.; Millan, M. Anthracene aquacracking using NiMo/SiO$_2$ catalysts in supercritical water conditions. *Fuel* **2016**, *182*, 740–748. [CrossRef]
9. Cabrales-Navarro, F.A.; Pereira-Almao, P. Catalytic steam cracking of a deasphalted vacuum residue using a Ni/K ultradispersed catalyst. *Energy Fuels* **2017**, *31*, 3121–3131. [CrossRef]
10. Kukushkin, R.G.; Eletskii, P.M.; Zaikina, O.O.; Sosnin, G.A.; Bulavchenko, O.A.; Yakovlev, V.A. Studying the steam cracking of heavy oil over iron- and molybdenum-containing dispersed catalysts in a flow-type reactor. *Catal. Ind.* **2018**, *10*, 344–352. [CrossRef]
11. Clark, P.D.; Kirk, M.J. Studies on the upgrading of bituminous oils with water and transition metal catalysts. *Energy Fuels* **1994**, *8*, 380–387. [CrossRef]
12. Mironenko, O.O.; Sosnin, G.A.; Eletskii, P.M.; Gulyaeva, Y.K.; Bulavchenko, O.A.; Stonkus, O.A.; Rodina, V.O.; Yakovlev, V.A. A study of the catalytic steam cracking of heavy crude oil in the presence of a dispersed molybdenum-containing catalyst. *Pet. Chem.* **2017**, *57*, 618–629. [CrossRef]
13. Golmohammadi, M.; Ahmadi, S.J.; Towfighi, J. Catalytic cracking of heavy petroleum residue in supercritical water: Study on the effect of different metal oxide nanoparticles. *J. Supercrit. Fluids* **2016**, *113*, 136–143. [CrossRef]
14. Fedyaeva, O.N.; Antipenko, V.R.; Vostrikov, A.A. Conversion of sulfur-rich asphaltite in supercritical water and effect of metal additives. *J. Supercrit. Fluids* **2014**, *88*, 105–116. [CrossRef]

15. Fumoto, E.; Sato, S.; Takanohashi, T. Characterization of an Iron-Oxide-Based Catalyst Used for Catalytic Cracking of Heavy Oil with Steam. *Energy Fuels* **2018**, *32*, 2834–2839. [CrossRef]
16. Xu, L.; Liu, Y.; Li, Y.; Lin, Z.; Ma, X.; Zhang, Y.; Argyle, M.D.; Fan, M. Catalytic CH_4 reforming with CO_2 over activated carbon based catalysts. *Appl. Catal. A Gen.* **2014**, *469*, 387–397. [CrossRef]
17. Bermúdez, J.M.; Fidalgo, B.; Arenillas, A.; Menéndez, J.A. Dry reforming of coke oven gases over activated carbon to produce syngas for methanol synthesis. *Fuel* **2010**, *89*, 2897–2902. [CrossRef]
18. Bulushev, D.A.; Ross, J.R.H. Towards sustainable production of formic acid. *ChemSusChem* **2018**, *11*, 821–836. [CrossRef] [PubMed]
19. Bulushev, D.A.; Ross, J.R.H. Heterogeneous Catalysts for hydrogenation of CO_2 and bicarbonates to formic acid and formates. *Catal. Rev.* **2018**, *60*, 566–593. [CrossRef]
20. Reichert, J.; Brunner, B.; Jess, A.; Wasserscheid, P.; Albert, J. Biomass Oxidation to Formic Acid in Aqueous Media Using Polyoxometalate Catalysts—Boosting Fa Selectivity by in-Situ Extraction. *Energy Environ. Sci.* **2015**, *8*, 2985–2990. [CrossRef]
21. Cui, Q.; Zhou, Y.; Wei, Q.; Tao, X.; Yu, G.; Wang, Y.; Yang, J. Role of the Zeolite Crystallite Size on Hydrocracking of Vacuum Gas Oil over NiW/Y-ASA Catalysts. *Energy Fuels* **2012**, *26*, 4664–4670. [CrossRef]
22. Landau, M.V.; Vradman, L.; Valtchev, V.; Lezervant, J.; Liubich, E.; Talianker, M. Hydrocracking of Heavy Vacuum Gas Oil with a Pt/H-beta-Al_2O_3 Catalyst: Effect of Zeolite Crystal Size in the Nanoscale Range. *Ind. Eng. Chem. Res.* **2003**, *42*, 2773–2782. [CrossRef]
23. Camblor, M.A.; Corma, A.; Martinez, A.; Martinez-Soria, V.; Valencia, S. Mild Hydrocracking of Vacuum Gasoil over NiMo-Beta Zeolite Catalysts: The Role of the Location of the NiMo Phases and the Crystallite Size of the Zeolite. *J. Catal.* **1998**, *179*, 537–547. [CrossRef]
24. Dik, P.P.; Danilova, I.G.; Golubev, I.S.; Kazakov, M.O.; Nadeina, K.A.; Budukva, S.V.; Pereyma, V.Y.; Klimov, O.V.; Prosvirin, I.P.; Gerasimov, E.Y.; et al. Hydrocracking of vacuum gas oil over NiMo/zeolite-Al_2O_3: Influence of zeolite properties. *Fuel* **2019**, *237*, 178–190. [CrossRef]
25. Surovikin, V.F.; Plaxin, G.V.; Semikolenov, V.A.; Likholobov, V.A.; Tiunova, I.J. Porous Carbonaceous Material. U.S. Patent 4978649, 18 December 1990.
26. Yermakov, Y.I.; Surovikin, V.F.; Plaxin, G.V.; Semikolenov, V.A.; Likholobov, V.A.; Chuvilin, L.V.; Bogdanov, S.V. New carbon material as support for catalysts. *React. Kinet. Catal. Lett.* **1987**, *33*, 435–440. [CrossRef]
27. Wang, Q.; Tsumori, N.; Kitta, M.; Xu, Q. Fast Dehydrogenation of Formic Acid over Palladium Nanoparticles Immobilized in Nitrogen-Doped Hierarchically Porous Carbon. *ACS Catal.* **2018**, *8*, 12041–12045. [CrossRef]
28. Navlani-García, M.; Mori, K.; Salinas-Torres, D.; Kuwahara, Y.; Yamashita, H. New Approaches toward the Hydrogen Production from Formic Acid Dehydrogenation over Pd-Based Heterogeneous Catalysts. *Front. Mat.* **2019**, *6*. [CrossRef]
29. Podyacheva, O.; Bulushev, D.; Suboch, A.; Svintsitskiy, D.; Lisitsyn, A.; Modin, E.; Chuvilin, A.; Gerasimov, E.; Sobolev, V.; Parmon, V. Highly Stable Single-Atom Catalyst with Ionic Pd Active Sites Supported on N-Doped Carbon Nanotubes for Formic Acid Decomposition. *ChemSusChem* **2018**, *11*, 3724–3727. [CrossRef]
30. Bulushev, D.A.; Zacharska, M.; Shlyakhova, E.V.; Chuvilin, A.L.; Guo, Y.; Beloshapkin, S.; Okotrub, A.V.; Bulusheva, L.G. Single Isolated Pd^{2+} Cations Supported on N-Doped Carbon as Active Sites for Hydrogen Production from Formic Acid Decomposition. *ACS Catal.* **2016**, *6*, 681–691. [CrossRef]
31. Fujitsuka, H.; Nakagawa, K.; Hanprerakriengkrai, S.; Nakagawa, H.; Tago, T. Hydrogen Production from formic acid using Pd/C, Pt/C, and Ni/C catalysts prepared from Ion-exchange resins. *J. Chem. Eng. Jpn.* **2019**, *52*, 423–429. [CrossRef]
32. Wang, S.; Yin, Q.; Guo, J.; Zhu, L. Influence of Ni Promotion on Liquid Hydrocarbon Fuel Production over Co/CNT Catalysts. *Energy Fuels* **2013**, *27*, 3961–3968. [CrossRef]
33. Satterfield, C.N. *Heterogeneous Catalysis in Industrial Practice*, 2nd ed.; McGraw-Hill: New York, NY, USA, 1991.
34. Escola, J.M.; Aguado, J.; Serrano, D.P.; Briones, L.; Díaz De Tuesta, J.L.; Calvo, R.; Fernandez, E. Conversion of polyethylene into transportation fuels by the combination of thermal cracking and catalytic hydroreforming over Ni-supported hierarchical beta zeolite. *Energy Fuels* **2012**, *26*, 3187–3195. [CrossRef]
35. Akhmedov, V.M.; Al-Khowaiter, S.H.; Akhmedov, E.; Sadikhov, A. Low temperature hydrocracking of hydrocarbons on Ni-supported catalysts. *Appl. Catal. A* **1999**, *181*, 51–61. [CrossRef]

36. Li, X.B.; Wang, S.R.; Cai, Q.J.; Zhu, L.J.; Yin, Q.Q.; Luo, Z.Y. Effects of preparation method on the performance of Ni/Al$_2$O$_3$ catalysts for hydrogen production by bio-oil steam reforming. *Appl. Biochem. Biotechnol.* **2012**, *168*, 10–20. [CrossRef] [PubMed]
37. De Haan, R.; Joorst, G.; Mokoena, E.; Nicolaides, C.P. Non-sulfided nickel supported on silicated alumina as catalyst for the hydrocracking of n-hexadecane and of iron-based Fischer–Tropsch wax. *Appl. Catal. A* **2007**, *327*, 247–254. [CrossRef]

© 2020 by the authors. Licensee MDPI, Basel, Switzerland. This article is an open access article distributed under the terms and conditions of the Creative Commons Attribution (CC BY) license (http://creativecommons.org/licenses/by/4.0/).

Article

Hydrogen Production through Autothermal Reforming of Ethanol: Enhancement of Ni Catalyst Performance via Promotion

Ekaterina Matus [1], Olga Sukhova [1], Ilyas Ismagilov [1,*], Mikhail Kerzhentsev [1], Olga Stonkus [1] and Zinfer Ismagilov [1,2]

[1] Boreskov Institute of Catalysis SB RAS, 630090 Novosibirsk, Russia; matus@catalysis.ru (E.M.); sukhova@catalysis.ru (O.S.); ma_k@catalysis.ru (M.K.); stonkus@catalysis.ru (O.S.); zinfer1@mail.ru (Z.I.)
[2] Federal State Budget Scientific Centre «The Federal Research Center of Coal and Coal-Chemistry of Siberian Branch of the Russian Academy of Sciences», 650000 Kemerovo, Russia
* Correspondence: iismagil@catalysis.ru

Abstract: Autothermal reforming of bioethanol (ATR of C_2H_5OH) over promoted $Ni/Ce_{0.8}La_{0.2}O_{1.9}$ catalysts was studied to develop carbon-neutral technologies for hydrogen production. The regulation of the functional properties of the catalysts was attained by adjusting their nanostructure and reducibility by introducing various types and content of M promoters (M = Pt, Pd, Rh, Re; molar ratio M/Ni = 0.003–0.012). The composition–characteristics–activity correlation was determined using catalyst testing in ATR of C_2H_5OH, thermal analysis, N_2 adsorption, X-ray diffraction, transmission electron microscopy, and EDX analysis. It was shown that the type and content of the promoter, as well as the preparation mode (combined or sequential impregnation methods), determine the redox properties of catalysts and influence the textural and structural characteristics of the samples. The reducibility of catalysts improves in the following sequence of promoters: Re < Rh < Pd < Pt, with an increase in their content, and when using the co-impregnation method. It was found that in ATR of C_2H_5OH over bimetallic $Ni-M/Ce_{0.8}La_{0.2}O_{1.9}$ catalysts at 600 °C, the hydrogen yield increased in the following row of promoters: Pt < Rh < Pd < Re at 100% conversion of ethanol. The introduction of M leads to the formation of a NiM alloy under reaction conditions and affects the resistance of the catalyst to oxidation, sintering, and coking. It was found that for enhancing Ni catalyst performance in H_2 production through ATR of C_2H_5OH, the most effective promotion is with Re: at 600 °C over the optimum $10Ni-0.4Re/Ce_{0.8}La_{0.2}O_{1.9}$ catalyst the highest hydrogen yield 65% was observed.

Keywords: renewable hydrogen; biofuel; reforming of bioethanol; bimetallic catalyst; modifier

Citation: Matus, E.; Sukhova, O.; Ismagilov, I.; Kerzhentsev, M.; Stonkus, O.; Ismagilov, Z. Hydrogen Production through Autothermal Reforming of Ethanol: Enhancement of Ni Catalyst Performance via Promotion. *Energies* **2021**, *14*, 5176. https://doi.org/10.3390/en14165176

Academic Editor: Luca Gonsalvi

Received: 30 June 2021
Accepted: 17 August 2021
Published: 21 August 2021

Publisher's Note: MDPI stays neutral with regard to jurisdictional claims in published maps and institutional affiliations.

Copyright: © 2021 by the authors. Licensee MDPI, Basel, Switzerland. This article is an open access article distributed under the terms and conditions of the Creative Commons Attribution (CC BY) license (https://creativecommons.org/licenses/by/4.0/).

1. Introduction

Currently, there is a rapid growth in the population of the Earth. Over the past 200 years, the number of people has grown by ~6 billion, reaching 7.8 billion in 2020. According to forecasts, by the end of the century, the world's population will reach 11 billion people [1,2]. To meet the demand of a growing population, the world needs more and more energy every year. Fossil fuels continue to be the main source of energy resources. Humanity consumes ~22 million tons of coal, ~12 million tons of oil, and ~10 billion m^3 of natural gas per day [3]. The constant growth in fossil fuel consumption is accompanied by an increase in the concentration of carbon dioxide in the atmosphere, which is the cause of climate change. To reduce the negative impacts on the environment and rational use of natural resources, it is urgent to develop technologies for the decarbonization of the energy system [4–6]. This is in line with the global strategy to reduce greenhouse gas emissions and the realization of the Paris Agreement's tasks to create a climate-neutral society by 2050 [7].

Under these conditions, hydrogen energy has already been recognized as a reasonable decision in the struggle for climate neutrality [8,9]. By 2050, it is expected that 24% of the world's energy needs will be met by hydrogen. The priority is to obtain renewable hydrogen, produced mainly with the use of wind and solar energy. In the short and medium term, low carbon footprint technologies will prevail, leading to reduced CO_2 emissions. Currently, the volume of hydrogen production in the world is estimated at 75 million tons per year and is expected to increase by 30% in the next five years [10]. Reforming natural gas is one of the main conventional ways of hydrogen production [11–15]. This is the least expensive and energy-efficient method, but to prevent CO_2 emissions, the use of carbon capture and storage technology is required [16,17]. It is advisable for the production of hydrogen to use biofuels (biogas, bioethanol), the source of which can be a renewable raw material–biomass [18–24].

Bioethanol takes the top place in the list of liquid biofuels. World bioethanol production in 2020 amounted to 26 billion gallons [25]. The world leader in the bioethanol production is the United States, which generated about 13.8 billion gallons. The second-largest producer country is Brazil, which produced 7.9 billion gallons of ethanol. The bulk of bioethanol is derived from corn and sugar cane. Various agricultural crops with a high starch or sugar content can also become raw materials for the bioethanol production, for example, cassava, potatoes, sugar beets, sweet potatoes, sorghum, barley, etc. The raw materials can also be various agricultural and forestry waste: wheat straw, rice straw, sugarcane bagasse, sawdust. The growth of biomass is accompanied by the consumption of atmospheric CO_2. Thus, using bioethanol as a feedstock in H_2 production reduces the consumption of fossil fuels and provides carbon neutrality of technology.

The most efficient process for producing hydrogen from ethanol is autothermal reforming (ATP of C_2H_5OH) [26]:

$$CH_3CH_2OH + 1.8H_2O + 0.6O_2 \longrightarrow 4.8H_2 + 2CO_2 \quad \Delta H°_{298} = +4.4 \text{ kJ/mol}$$

The energy neutrality of this reaction makes it possible to refer it to energy-saving processes. In addition to the favorable energy balance, this process is characterized by a high yield of H_2. The effective conversion of bioethanol requires the solution of important problems of increasing the activity of catalysts and their resistance to deactivation. The chemical formula and nanostructure of catalysts strongly affect H_2 yield and the composition of the reaction products in ATR of C_2H_5OH [27–30]. The mode of the first stages of ethanol transformation (dehydrogenation to acetaldehyde or dehydration to ethylene) depends mainly on the properties of the support [26]. Diffusion and transformation of C_2-intermediates are controlled by the metal-support interface, while the decomposition of C_2-intermediates and the conversion of C_1-reaction products occur with the participation of metal centers of the active component. The noble metals Rh, Ag, Au, Pd, Pt, Ru, Re, as well as Ni, or Co, are used as an active metal, while for stabilization of their highly dispersed forms, various oxides (SiO_2, MgO, La_2O_3, CeO_2-Al_2O_3, CeO_2-La_2O_3, $CeMnO_2$, $MgAl_2O_4$) are applied as support [31–35].

Cerium dioxide is a suitable support for ATR of C_2H_5OH catalysts due to its redox properties, high oxygen capacity, and the possibility of realization of the strong metal-support interaction [36–40]. The intrinsic catalytic activity of CeO_2-based supports and the degree of their interaction with the active component can be regulated by its doping. It was shown that among the tested dopants (M = Gd, La, Mg), La has a more pronounced positive effect on the state and functionality of $Ni/Ce_{1-x}M_xO_y$ (M = Gd, La, Mg, x = 0–0.9, $1.5 \leq y \leq 2.0$) [41–44]. The introduction of La as a modifier in the support composition enhances the metal-support interaction, which improves Ni dispersion and catalyst stability under the ATR of C_2H_5OH. However, it also leads to a diminution of Ni^{n+} reducibility that can decrease the concentration of Ni^0 active sites and, consequently, H_2 yield.

The introduction of a promoter is a rather wide-spread approach directed to the improvement of functional characteristics of Ni catalysts developed for various catalytic processes [45–49]. It is shown that bimetallic catalysts have advantages over monometallic ones in the reforming of C_2H_5OH [31,35,50–53]. In particular, it is indicated [35] that for $Ni/CeMnO_2$, the introduction of Cu or Fe increased the ethanol conversion from 57 to

70% and 61% consequently. Ni-Fe/CeMnO$_2$ catalyst provides higher hydrogen yield (60%) among the other samples because of conducting the reaction to dehydrogenation route, while the feature of Ni-Co/CeMnO$_2$ sample was high CO selectivity due to the impact of Co in the progress of water–gas shift reaction. It is noted that an increase of H$_2$ yield over Ni-Co/Ce-Zr-O and Ni-Pd/SiO$_2$ catalysts is correlated with an increase of reducibility of the bimetallic sample in comparison with appropriate monometallic Ni catalyst [31,50]. The Ru or Rh additives stabilized Co metallic phase under operation in oxidative conditions [54]. A study of a series of Cu-Ni/SiO$_2$ catalysts with different Cu/Ni molar ratios showed that the Cu-rich catalysts had a higher resistance of coking [55]. Thus, the mechanism of action of the promoter is specific and strongly depends on the composition and method of preparation of the catalyst. It is possible to increase the reducibility of the active component, improve its dispersion or resistance to coking.

This work is devoted to the development of effective bimetallic catalysts for hydrogen production through ATR of C$_2$H$_5$OH. To improve the functional properties of the Ni/Ce$_{0.8}$La$_{0.2}$O$_{1.9}$ catalysts, their nanostructure and reducibility were regulated by introducing various types and content of M promoters (M = Pt, Pd, Rh, Re; molar ratio M/Ni = 0.003–0.012).

2. Materials and Methods

The Ni-M/Ce$_{0.8}$La$_{0.2}$O$_{1.9}$ catalysts (M = Pt, Pd, Rh, and Re; molar ratio M/Ni = 0.003–0.012) were prepared by the combined incipient wetness impregnation method unless otherwise specified. For this, Ce$_{0.8}$La$_{0.2}$O$_{1.9}$ support was impregnated by an aqueous solution of the mixture (Ni + M) of metal precursors with a specified concentration. The description of the preparation mode and properties of the Ce$_{0.8}$La$_{0.2}$O$_{1.9}$ support can be found in our previous paper [56]. After the impregnation, the Ni-M/Ce$_{0.8}$La$_{0.2}$O$_{1.9}$ catalysts were dried at 90 °C for 6 h, calcined at 500 °C for 4 h in the air. The Ni(NO$_3$)$_2$·6H$_2$O was used as Ni precursor, while H$_2$PtCl$_6$·6H$_2$O, Pd(NO$_3$)$_2$, RhCl$_3$·3H$_2$O, or NH$_4$ReO$_4$ compounds were used as precursors for Pt, Pd, Rh, or Re promoters, respectively.

In some specially stipulated cases, the catalysts were obtained by the sequential incipient wetness impregnation method. In this case, the Ce$_{0.8}$La$_{0.2}$O$_{1.9}$ support was impregnated by an aqueous solution of Ni(NO$_3$)$_2$·6H$_2$O with a given concentration. After the impregnation, the Ni/Ce$_{0.8}$La$_{0.2}$O$_{1.9}$ catalyst was dried at 90 °C for 6 h, calcined at 500 °C for 4 h in the air, and then impregnated by an aqueous solution of a promoter (M) precursor with a specified concentration. Then Ni-M/Ce$_{0.8}$La$_{0.2}$O$_{1.9}$ samples were dried at 90 °C for 6 h and calcined at 500 °C for 4 h in the air.

The Ni content was equal to 10 wt.%, while the molar ratio M/Ni was 0.003 or 0.012. The samples are noted according to their composition and synthesis procedure: the number means the molar ratio M/Ni, while "C" and "S" correspond to the samples obtained by combined and sequential impregnation, respectively.

The catalysts were thoroughly studied by X-ray fluorescence spectroscopy, thermal analysis (TA) (thermogravimetric (TG), differential thermogravimetric (DTG), and differential thermal analysis (DTA)), N$_2$ adsorption, X-ray diffraction, transmission electron microscopy, and EDX analysis. A description of devices and conditions for studying materials by physicochemical methods can be found in our earlier publications [41,44,47].

ATR of C$_2$H$_5$OH was investigated in a flow setup with a quartz reactor (14 mm i.d.) at atmospheric pressure, temperature 200–700 °C, a flow rate of 230 mL/min and the molar ratio between reagents C$_2$H$_5$OH:H$_2$O:O$_2$:He = 1:3:0.5:1 according to the method described in [41]. Note that the influence of the reaction conditions (500–700 °C, C$_2$H$_5$OH:H$_2$O = 1–4, C$_2$H$_5$OH:O$_2$ = 0.2–0.8) was preliminarily studied, and the optimal molar ratio of the reagents for the maximum hydrogen yield was selected. In contrast to our previous studies [41,57], there was no reduction of catalysts before catalytic activity tests. In this case, the active centers will be formed directly under the reaction conditions. The ability to self-activate in the reaction environment ensures that the catalyst can operate on a daily start-up and shut-down cycle without requiring activation before use [58]. It is essential for hydrogen production through reforming for fuel cell technology [59,60].

3. Results and Discussion

3.1. Characteristics of the Ni-M/Ce$_{0.8}$La$_{0.2}$O$_{1.9}$

The Ni-M/Ce$_{0.8}$La$_{0.2}$O$_{1.9}$ samples were prepared by the incipient wetness impregnation method, in which precursors of the active component are introduced into the support matrix followed by thermal treatment. The characteristics of the decomposition of metal precursors and the formation of catalysts have been studied by thermal analysis. Figure 1 demonstrates typical TG, DTG, and DTA curves of dried unpromoted and promoted Ni catalysts.

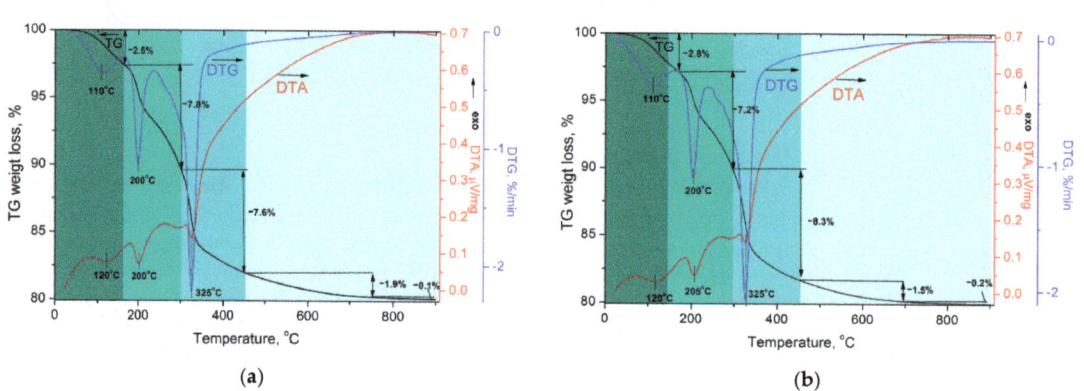

Figure 1. Thermal analyses in the air of dried Ni (a) and Ni-Pt-0.012 (C) (b) catalysts.

For Ni catalyst in a low-temperature region (T < 150 °C), an endothermic effect at T_{DTA} = 125 °C is observed. It is accompanied by a weight loss ($-\Delta m/m$ = 2.5%) through water desorption. In the temperature range 150–350 °C, there are two endothermic effects at T_{DTA} = 200 and 325 °C. The weight losses of 7.8 and 7.6% were connected with the decomposition of nitrate nickel hydrate to anhydrous Ni(NO$_3$)$_2$ and then transformation of Ni(NO$_3$)$_2$ to oxide NiO, correspondingly [61]. At a temperature of 450–900 °C, the change in the weight of the sample is apparently associated with dehydroxylation of the support surface. The total weight loss is equal to 17.4%. Similar behavior was observed for all other Ni-M/Ce$_{0.8}$La$_{0.2}$O$_{1.9}$ catalysts. As an example, on the derivatogram of the Ni-Pt/Ce$_{0.8}$La$_{0.2}$O$_{1.9}$ sample, there are no effects corresponding to the decomposition of the Pt precursor due to its low content (Figure 1b).

From N$_2$ adsorption data (Figure 2), it follows that fresh Ni-M catalysts are mesoporous materials: the type IV adsorption isotherms with a hysteresis loop of type H3 are observed, which usually indicates the pore shape is wedged with the opening at both ends or groove pores are formed by flaky particles [62]. Hysteresis at partial pressure P/Po = 0.7–1.0 corresponds to the presence of texture mesoporosity [39]. The samples show bimodal pore size distribution with a maximum at ca. 4 and 18 nm (Figure 2c). The texture characteristics of the samples are weakly dependent on the type of promoter: S_{BET} = 70 ± 5 m^2/g, V_{pore} = 0.20 ± 0.01 cm^3/g, and D_{pore} = 11.5 ± 0.9 nm (Table 1), which are typical values for materials of such composition [32,35,63]. When using the sequential impregnation method, there is a tendency to some decrease in specific surface area (69 → 56 m^2/g). This is probably due to differences in the heat treatment procedure for these samples. Single calcination at 500 °C instead of two times reduces the degree of sintering of the material. Note that the specific surface area of the catalysts is 25% lower than the S_{BET} of the $Ce_{0.8}La_{0.2}O_{1.9}$ support (94 m^2/g), which is due to the partial jamming of the support pores by particles of the active component.

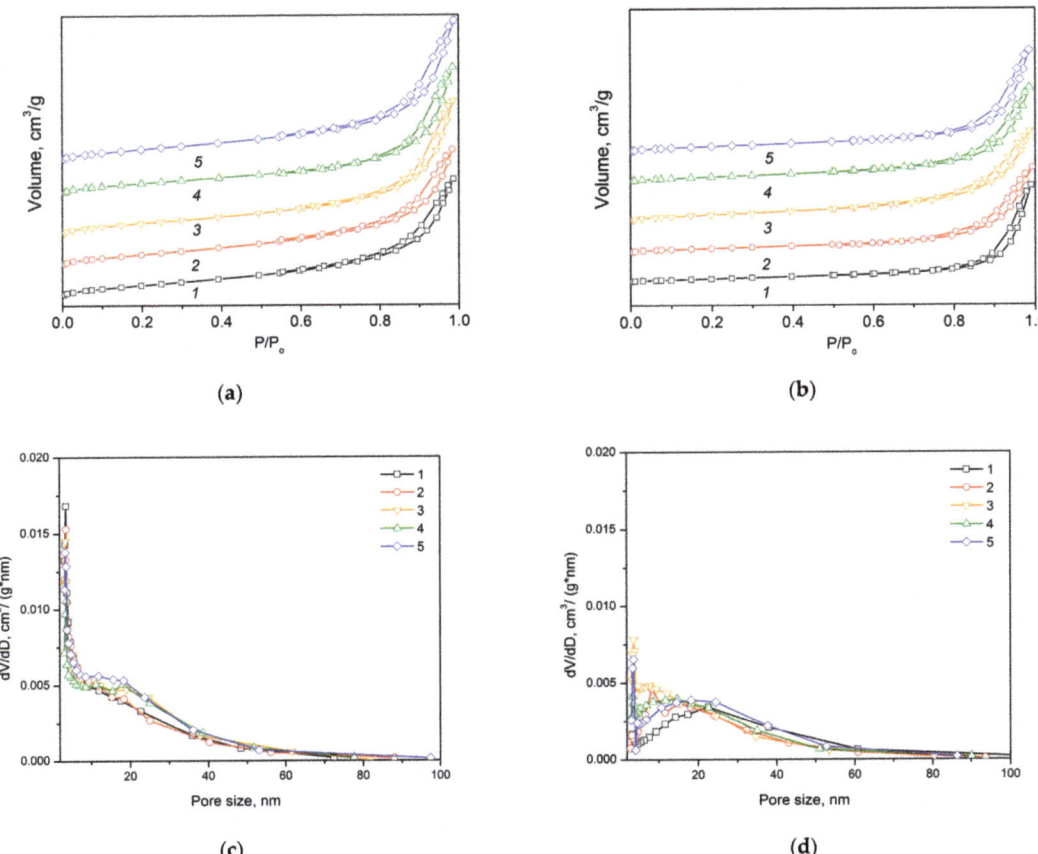

Figure 2. N$_2$ adsorption–desorption isotherms (**a**,**b**) and pore size distributions (**c**,**d**) for fresh (**a**,**c**) and spent (**b**,**d**) catalysts.

1-Ni; 2-Ni-Pt-0.012 (C); 3-Ni-Pd-0.012 (C); 4-Ni-Rh-0.012 (C); 5-Ni-Re-0.012 (C).

XRD patterns of the fresh catalysts are depicted in Figure 3. The pattern of fresh catalysts shows peaks characteristic of the CeO$_2$-based phase of support and NiO of the active component. The parameter of the unit cell (*a*) of CeO$_2$-based phase is 5.478 Å that is higher than *a* of undoped CeO$_2$ (5.411 Å, JCPDS-34-394). This confirms the presence of a CeO$_2$-based solid solution in which some of the Ce^{4+} cations are replaced by La^{3+} cations with a larger ionic radius (0.116 vs. 0.097 nm).

The coherent scattering region (CSR) was calculated by the Selyakov–Scherrer method from the broadening of the diffraction peak 1.1.1 assigned to phases having a cubic structure of the fluorite type; CSR (NiO)–peak 2.0.0 NiO phases; CSR (Ni)–peak 2.0.0 phases of Ni°. The average crystallite size of CeO$_2$-based solid solution is equal to 8.0 nm, and it is the same for all samples. Thus, the phase composition of the support and its structural characteristics do not change as a result of the introduction of Ni or Ni-M active component (Table 1). The average crystallite size of the NiO phase is slightly smaller for Ni, Ni-Pd, and Ni-Rt catalysts in comparison to Ni-Pt and Ni-Re samples (16.5 vs. 19.0 nm), which indicates some effect of the second metal. The diffraction patterns of the Ni-M samples show no peaks related to the corresponding M-containing phases because of the low content of M (less than 1 wt.%) and its high-dispersed state (Figure 3).

TEM data (Figure 4) confirms the XRD study results. On the surface of CeO_2-based solid solution with crystallite size no greater than 10 nm, NiO particles of 10–20 nm in size were observed.

Table 1. Textural and structural characteristics of the Ni-M catalysts.

Sample [1]		Textural Characteristics				Structural Characteristics	
		S_{BET}, m^2/g	V_{pore}, cm^3/g	D_{pore}, nm	Phase Composition	CSR (nm)/Parameter of the Unit Cell (Å) for	
						CeO_2-Based Phase	Ni-Containing Phase
$Ce_{0.8}La_{0.2}O_{1.9}$		94	0.19	7.9	CeO_2	8.0/5.478	-
Ni	F	69	0.19	10.8	CeO_2, NiO	8.0/5.478	16.5
	S	24	0.15	25.7	CeO_2, Ni°	18.0/5.479	20.0/3.525
Ni-Pt-0.012 (C)	F	70	0.19	10.7	CeO_2, NiO	8.0/5.480	16.5
	S	26	0.14	21.0	CeO_2, Ni°	14.0/5.481	20.0/3.526
Ni-Pd-0.012 (C)	F	69	0.21	12.3	CeO_2, NiO	8.0/5.480	17.0
	S	41	0.15	14.5	CeO_2, NiPd	14.0/5.483	20.0/3.532
Ni-Rh-0.012 (C)	F	64	0.20	12.4	CeO_2, NiO	8.0/5.480	19.0
	S	34	0.15	18.0	CeO_2, NiRh	14.0/5.483	20.0/3.528
Ni-Re-0.012 (C)	F	75	0.22	11.9	CeO_2, NiO	8.0/5.480	19.0
	S	38	0.17	17.3	CeO_2, NiRe	13.0/5.488	16.0/3.538
Ni-Pd-0.003 (S)	F	62	0.19	12.0	CeO_2, NiO	8.0 /5.480	16.5
	S	43	0.18	16.6	CeO_2, NiPd	15.0/5.482	20.0/3.529
Ni-Pd-0.012 (S)	F	56	0.17	12.3	CeO_2, NiO	8.0/5.480	16.5
	S	60	0.16	11.1	CeO_2, NiPd	15.0/5.482	20.0/3.532

[1] F–fresh catalysts (before ATR of C_2H_5OH), S–spent catalyst (after ATR of C_2H_5OH).

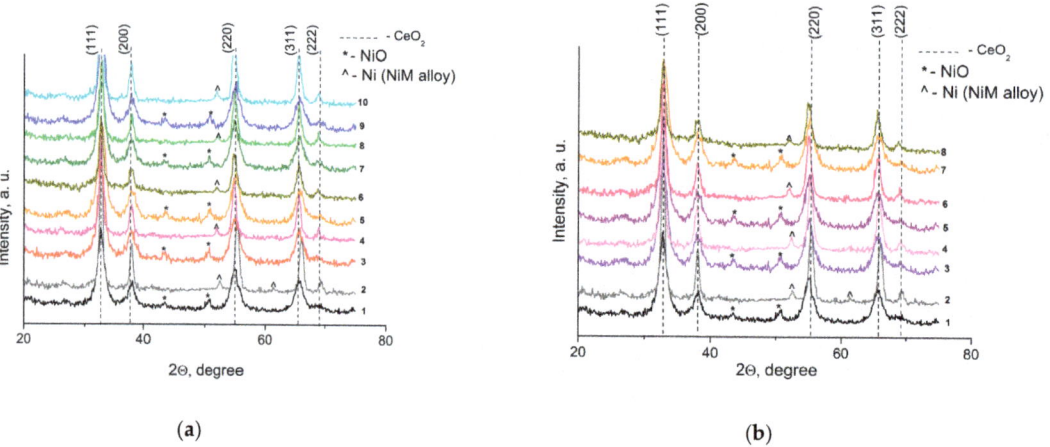

Figure 3. XRD patterns of fresh (odd numbers) and spent (even numbers) catalysts: influence of promoter type (a), promoter content and synthesis method (b). (a) 1,2—Ni; 3,4—Ni-Pt-0.012 (C); 5,6—Ni-Pd-0.012 (C); 7,8—Ni-Rh-0.012 (C); 9,10—Ni-Re-0.012 (C). (b) 1,2—Ni; 3,4—Ni-Pd-0.003 (S); 5,6—Ni-Pd-0.012 (S); 7,8—Ni-Pd-0.012 (C).

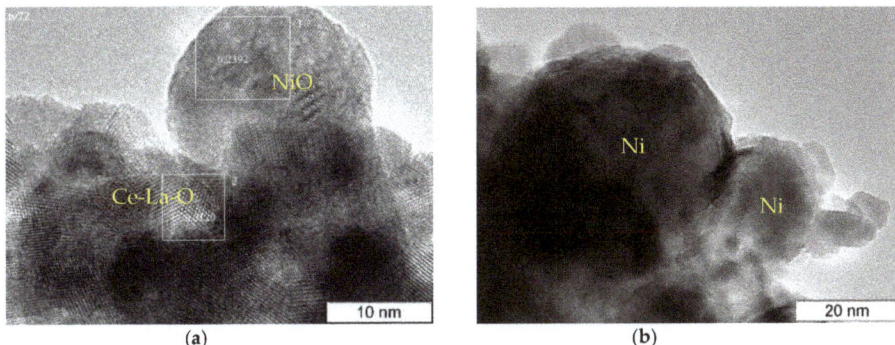

Figure 4. TEM images of the fresh (**a**) and spent (**b**) Ni catalyst.

In the case of catalyst Ni-Pd-0.012 (C), we studied the local distribution of constituent elements using the EDX analysis. Figure 5 shows a HAADF-STEM image of the fresh Ni-Pd-0.012 (C) catalyst with the corresponding EDX maps. It was seen that Ce and La had almost homogeneous distribution in the studied region, which confirmed the formation of a solid $Ce_xLa_{1-x}O_2$ solution. The opposite situation was observed for nickel: areas with an increased nickel concentration were well seen, which indicated the formation of its separate phase. It agreed well with the XRD data (Table 1). The palladium distribution map did not contain any areas of high concentration. This indicates the absence of individual palladium-containing nanoparticles in the studied area. However, due to the low palladium content, the PdL signal is very weak. Thus, the obtained data do not allow us to conclude whether palladium is located predominantly in the composition of Ni-rich particles or is localized on the surface of the fluorite phase.

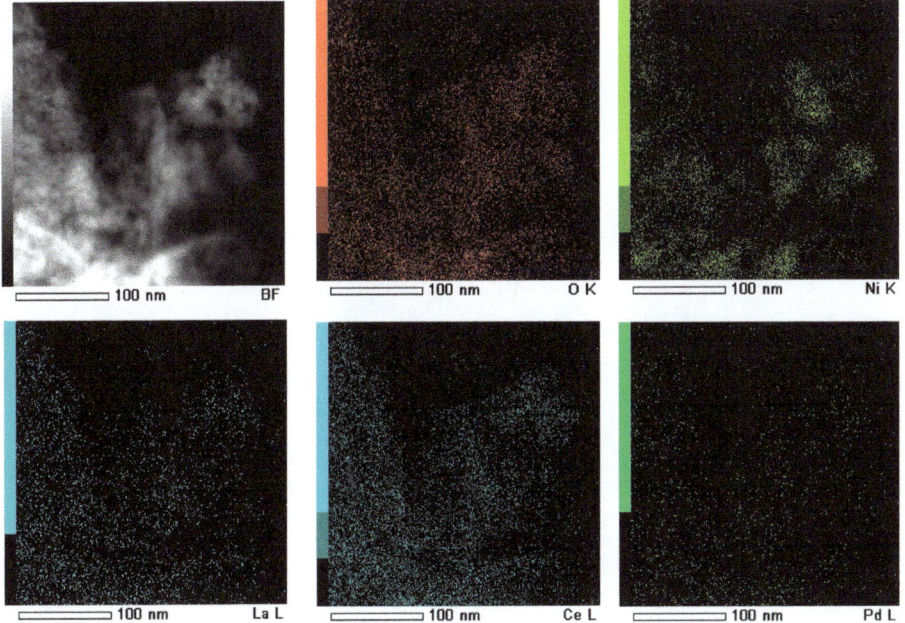

Figure 5. HAADF-STEM image of the fresh Ni-Pd-0.012 (C) catalyst and corresponding EDX mapping patterns.

The redox properties of the active component are an important characteristic of the reforming catalyst since they determine the concentration of active Ni° centers under the reaction conditions. Typically, the samples were activated by treatment in a hydrogen-containing mixture at a high temperature before the ATR of C_2H_5OH reaction [31,33,35]. This allows nickel oxide to be reduced and Ni° active sites to form. The Ni° state should be preserved under reaction conditions, while re-oxidation of Ni° will lead to catalyst deactivation. A positive quality of reforming catalytic systems is their ability to self-activate under reaction conditions [47,48]. In this case, no additional stage of activation was required, there was no consumption of H_2, and active centers were formed directly under the reaction conditions. Thus, to study the effect of the composition of Ni-M catalyst and the method of its synthesis on the reducibility of Ni^{2+} cations, the samples were studied by the thermal analysis in H_2/He (Figure 6). There were four temperature regions of weight loss which could be connected with water desorption (T < 200 °C), reduction of nickel oxide (300 °C < T < 600 °C), and reduction surface (200 °C < T < 300 °C) and bulk (600 °C < T < 900 °C) cerium dioxide. For the unpromoted Ni catalyst, three temperature peaks are observed during the reduction of nickel oxide species: at 405, 475, and 545 °C. The observed wide reduction region indicates the presence of various forms of Ni^{2+} stabilization in the support matrix. It is known [64] that reduction of large particles of NiO, characterized by weak metal-support interaction, occurs in a low-temperature area (T < 500 °C), while highly dispersed particles of NiO, characterized by strong metal-support interaction, is in the high-temperature region (T > 500 °C). It was also shown that the reduction of Ni^{2+} cations moved to a high-temperature region when the dispersion of Ni-supported nanoparticles and their sintering stability increase due to a decrease of the crystallite size of support [41].

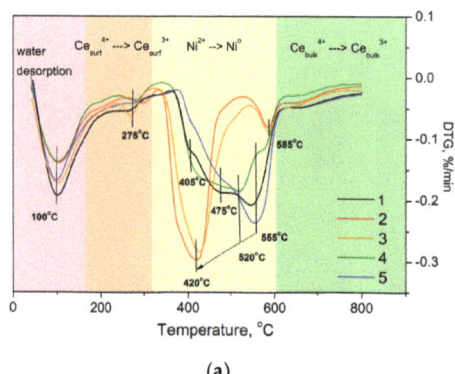
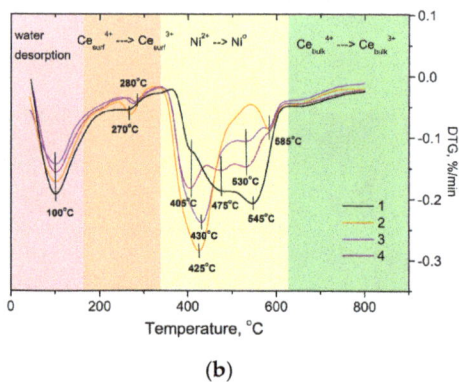

(a)   (b)

Figure 6. Thermal analyses in H_2/He for fresh Ni-M catalysts: effect of promoter type (**a**) and preparation mode (**b**). (**a**) 1—Ni; 2—Ni-Pt-0.012 (C); 3—Ni-Pd-0.012 (C); 4—Ni-Rh-0.012 (C); 5—Ni-Re-0.012 (C). (**b**) 1—Ni; 2—Ni-Pd-0.012 (C); 3—Ni-Pd-0.012 (S); 4—Ni-Pd-0.003 (S).

With the introduction of a promoter, the behavior of the Ni^{2+} reduction was changed (Figure 6). In particular, in the presence of Pt or Pd, the reduction of Ni^{2+} shifted to the low-temperature region with a maximum at ~420 °C that could be explained by the H_2 spillover effect [46,58]. This effect was more pronounced when higher content of promoter or the co-impregnation method of synthesis in contrast to the sequential impregnation method was used (Figure 6b). In the case of Rh, the reduction region remained the same, but the part of difficult-to-reduce Ni^{2+} species decreased. With the introduction of Re, on the contrary, the part of difficult-to-reduce Ni species increased, and the temperature maximum became equal to 545 °C. Note that due to the low content of promoters, their contribution to the reduction can be ignored in comparison with the reduction peak of Ni^{2+}.

As a rule, the reduction of Pt, Pd, and Rh species occurs at T < 200 °C, while reduction of Re cations takes place in the same region where Ni^{2+} cations are reduced [58,65]. Different reducibility of Ni-Re catalysts can be explained by the Ni–Re alloy formation under the conditions of TA in H_2/He that changes the kinetics of the active component reduction [66].

The degree of the interaction between metals in the composition of bimetal catalysts is rather different, and it depends on the chemical composition of the catalyst, its preparation method, conditions of its activation, and exploitation [46,67]. In the case of the weak interaction between nickel and a promoter, the formation of monometal particles took place; in the case of the strong interaction, a surface or bulk alloy was formed. It was shown [58] that for the Ni-Pd catalyst, the co-impregnation promoted the formation of bimetallic particles with a more uniform Pd distribution and a lower surface Pd concentration. In the case of sequential impregnation, the formation of bimetallic particles with a high surface Pd concentration–Pd clusters on the surface of Ni particles took place.

Thus, the type of promoter practically had no effect on the textural and structural properties of fresh Ni-M catalysts. On the contrary, the introduction of even a small amount of a promoter changes the reducibility of the Ni active component. The reducibility of catalysts improves in the following sequence of promoters Re < Rh < Pd < Pt, with an increase in their content, and when using the co-impregnation method. It is expected that the ability of a catalyst to form dispersed Ni^0 phase and retain it under reaction conditions will control the functional properties of catalysts in ATR of C_2H_5OH.

3.2. Activity of Ni-M/$Ce_{0.8}La_{0.2}O_{1.9}$ Catalysts in ATR of C_2H_5OH

Figure 7a shows a typical dependence of the conversion of ethanol and the yield of the reaction products in ATP of C_2H_5OH on temperature. It can be seen that in the presence of the Ni catalyst, the conversion of ethanol to a hydrogen-containing gas increased with an increase in the reaction temperature, reached 100% at 400 °C, and then did not change (Figure 7a). The main reaction products are H_2, CO, CO_2, and CH_4. Ethylene, acetaldehyde, and acetone were present in the reaction products in trace amounts only at T = 300 °C. With an increase in the reaction temperature from 200 to 700 °C, the hydrogen yield increased from 0 to ~43%. The yield of methane (Y_{CH4}) reached a maximum at a reaction temperature of 400 °C, and then at an increase in the reaction temperature decreased and became less than 1% at a T = 700 °C. The CO yield (Y_{CO}) increased over the temperature range 400–700 °C, and was equal to ~40% at 700 °C. The behavior of the temperature dependence of Y_{CH4} and Y_{CO} indicated an increase in the contribution of the reaction of steam conversion of methane with increasing temperature. The curve of the dependence of the CO_2 yield (Y_{CO2}) on temperature passed through a maximum (~67%) at a reaction temperature of 400–500 °C.

The temperature dependence of the activity indices of bimetallic Ni-M catalysts, in general, is similar to those observed in the Ni sample. The advantage of Ni-M systems is a lower temperature to achieve 100% conversion of ethanol and higher values of hydrogen yield (Figure 7, Table 2). Note that an increase in the promoter content has a positive effect on the process performance, while the method of introducing the metal does not matter (Table 2).

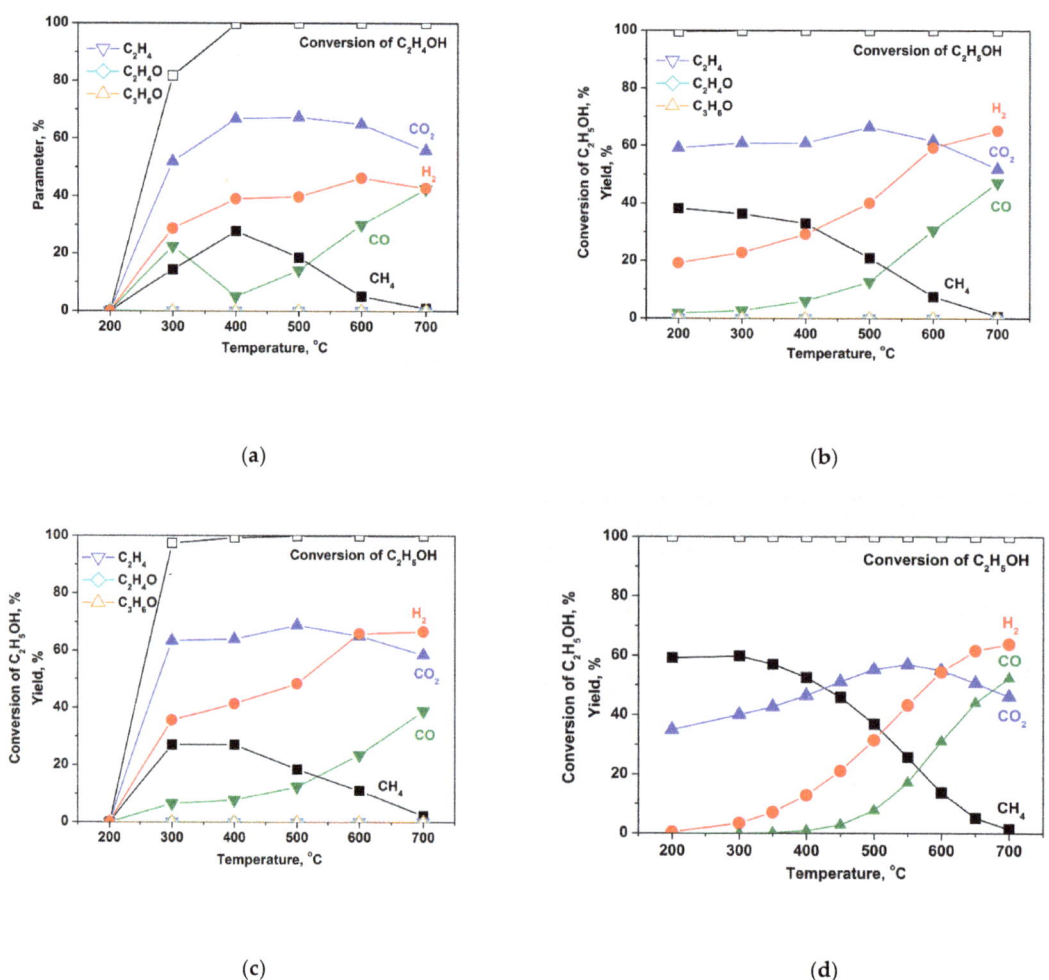

Figure 7. Temperature dependences of C_2H_5OH conversion and product yield in ATR of C_2H_5OH over Ni (**a**), Ni-Pd-0.012 (C) (**b**), Ni-Re-0.012 (**c**) catalysts, and thermodynamic equilibrium values (**d**).

Table 2. Activity of Ni-M catalysts in ATR of C_2H_5OH at 600 °C [1].

Catalyst	H_2 Yield, %	Selectivity, %		
		CO	CO_2	CH_4
Ni	46	30	65	5
Ni-Pt-0.012 (C)	51	27	66	7
Ni-Pd-0.012 (C)	59	31	62	7
Ni-Rh-0.012 (C)	54	30	64	6
Ni-Re-0.012 (C)	65	24	65	11
Ni-Pd-0.003 (S)	50	22	66	12
Ni-Pd-0.012 (S)	58	22	67	11

[1] Conversion of C_2H_5OH was equal to 100%.

At 600 °C, the hydrogen yield increased in the next row of promoters Pt < Rh < Pd < Re at 100% conversion of ethanol (Figure 8). For the Ni-Re sample, the obtained parameters of ATR reaction were close to thermodynamic equilibrium values, and the maximum H_2 yield (65%) was attained (Figures 7 and 8).

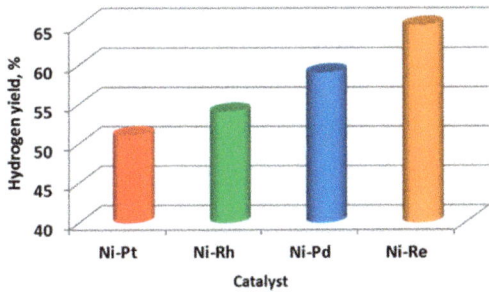

Figure 8. Hydrogen yield in ATR of C_2H_5OH at 600 °C over Ni-M-0.012 (C) catalyst: effect promoter type.

The study of catalyst characteristics after ATR of C_2H_5OH reaction shows that catalysts change their textural and structural properties. The values of S_{BET} decreased by 2–3 times, while the average pore diameter increases (Table 1). The behavior of the distribution of mesopores in size changed: the portion of large pores increased (Figure 2). It is associated with the intensification of sintering processes under reaction conditions. The increase in the contribution to the porosity of large interparticle pores was observed, which was indicated by the shift of the position of the hysteresis loop to the region of higher values of relative pressure (Figure 2). There is a tendency to increase the activity of catalysts with an increase in their resistance to sintering and retention of textural properties (Tables 1 and 2).

According to XRD data (Figure 2, Table 1), support preserves their phase composition and changes in the phase composition of catalysts are largely associated with changes in the structure of Ni-containing phases. It can be seen that for spent samples in comparison with fresh catalysts, the cell parameter of CeO_2-based solid solution practically does not change, but the average crystallite size increases (8 → 13–18 nm) which is associated with the intensification of sintering processes under the reaction conditions, wherein Ni sample is characterized by lower resistance to sintering, and the best is Ni-Re (13 vs. 18 nm). For all samples after the reaction a metallic nickel phase was formed, the average crystallite size was 16–20 nm (Figure 4b). There was no direct correlation between activity and reducibility (Figures 6 and 8) in contrast to the literature data [31,50]. All studied Ni and Ni-M catalysts have the capability to self-activation under reaction conditions and can be used without preliminary reduction.

It is noted that the value of the unit cell parameter (*a*) of Ni° in Ni-M catalysts (M = Pd, Rh, or Re) is larger than Ni° reference data (*a* = 3.523 Å): 3.532 Å (M = Pd), 3.528 Å (M = Rh), and 3.538 Å (M = Re). It is for these catalysts, in contrast to the Ni-Pt sample, that a more significant promotion effect is observed. It indicates the formation of Ni-M alloy that can change the detailed structure of the nanoparticle surface as well as modify electronic metal properties, which affects the activity of the catalysts due to the synergistic effect between metals.

According to the thermal analysis of spent catalysts in air, carbonaceous deposits were formed during the reaction. Their content depended on the catalyst composition and increased in the following row of samples: Ni-Pt (0.3%) < Ni ~ Ni-Rh (1.3%) < Ni-Re (4.5%) < Ni-Pd (7.5%). An increase in the activity was accompanied by an increase in the yield of carbonaceous by-products. Thus, a catalysts regeneration procedure should be developed.

It was mentioned above that for the enhancement of Ni catalyst performance in H_2 production through ATR of C_2H_5OH, the most effective promotion was with Re (Figure 8). This sample was distinguished by resistance to sintering, reducibility in the high-temperature region, and the formation of Ni-Re alloy particles which optimized the functional properties of the catalyst due to the synergistic effect between the metals. The developed catalysts provide equilibrium values of ethanol conversion 100% and hydrogen yield 65%, which is comparable or higher than those described in the literature (Table 3) and indicates that they are promising for use in hydrogen energy.

Table 3. Characteristics of the ATP C_2H_5OH process.

Catalyst	Process Conditions	H_2 Yield, mol H_2/mol C_2H_5OH	Reference
30Ni-1Rh/$Ce_{0.5}Zr_{0.5}O_2$	$C_2H_5OH:H_2O:O_2:He = 1:9:0.35:0$ T = 600 °C.	4.6	[68]
10Ni/ZrO_2/Al_2O_3	$C_2H_5OH:H_2O:O_2:N_2 = 1:6:0:24.5$ T = 500 °C.	4.1	[69]
10Ni-0.4Re/$Ce_{0.8}La_{0.2}O_{1.9}$	$C_2H_5OH:H_2O:O_2:He = 1:3:0.5:1$ T = 600 °C.	4.0	This work
10Co/MgO-Al_2O_3	$C_2H_5OH:H_2O:O_2:He = 1:3:0.4:0$ T = 600 °C.	3.8	[70]
Co/Pr/MgO-Al_2O_3	$C_2H_5OH:H_2O:O_2:He = 1:3:0.4:0$ T = 550 °C.	3.4	[71]
5Ni0.3Pt/10CeO_2/Al_2O_3	$C_2H_5OH:H_2O:O_2:He = 1:8:0.5:0$ T = 650 °C.	3.2	[72]
0.25Rh0.25Pt/ZrO_2	$C_2H_5OH:H_2O:O_2:He = 1:2:0.2:0$ T = 700 °C.	3.1	[73]
$LaNiFeO_3$	$C_2H_5OH:H_2O:O_2:He = 1:3:0.5:0$ T = 650 °C.	3.0	[74]
10Ni-3Pt/30CeO_2/SiO_2	$C_2H_5OH:H_2O:O_2:He = 1:3:0:0$ T = 750 °C.	2.4	[75]
2Ir/CeO_2	$C_2H_5OH:H_2O:O_2:He = 1:1.8:0.6:0$ T = 700 °C.	2.2	[76]

4. Conclusions

An important issue of our time is the development of technologies for the decarbonization of the energy system. This will reduce the greenhouse effect and become the key to the sustainable development of society. A promising carbon-neutral technology is a production of hydrogen from ethanol, the source of which can be a renewable raw material, biomass. Hydrogen is an important reagent and an alternative energy carrier with high ecological properties. In this work, the development of efficient bimetallic catalysts for the production of hydrogen by ATR of C_2H_5OH was carried out.

A series of Ni-M/$Ce_{0.8}La_{0.2}O_{1.9}$ catalysts were prepared at the variation of type (M = Pt, Pd, Rh, and Re) and content (molar ratio M/Ni = 0.003–0.012) of a promoter. The genesis of materials and their properties were systematically studied by thermal analysis, X-ray fluorescence analysis, N_2 adsorption, XRD, and TEM. It was found that the prepared catalysts are mesoporous materials with analogous textural properties (S_{BET} = 70 ± 5 m^2/g, V_{pore} = 0.20 ± 0.01 cm^3/g, and D_{pore} = 11.5 ± 0.9 nm) and phase composition (NiO with an average particle size of 18 ± 1 nm, Ce-La-O solid solution with an average crystallite size of 8.0 nm). It was shown that the reducibility of Ni^{2+} cations is regulated by the type and content of promoter M as well as the mode of its introduction. It enhances in the following sequence of promoters Re < Rh < Pd < Pt, with an increase in their content, and when using the co-impregnation method. The effect of promoter on the functional properties of catalysts in ATR of C_2H_5OH was studied, and the optimal composition of the catalyst was selected. The Ni-M/$Ce_{0.8}La_{0.2}O_{1.9}$ catalysts have the ability to self-activation under the reaction conditions, which makes it possible to exclude the catalyst pre-reduction before the ATR of C_2H_5OH. The catalysts after reaction retain sufficient textural characteristics and dispersion of the active component. To a greater extent, this is observed when rhenium is used as a promoter. With the optimum catalyst 10Ni-0.4Re/$Ce_{0.8}La_{0.2}O_{1.9}$, the high hydrogen yield of 65% in ATR of C_2H_5OH was achieved.

Thus, the optimal composition of the catalyst and mode of its preparation were determined. Application of developed 10Ni-0.4Re/$Ce_{0.8}La_{0.2}O_{1.9}$ catalyst for autothermal bioethanol reforming reduces the fossil fuel consumption and provides carbon neutrality of H_2 producing technology.

Author Contributions: Conceptualization, M.K. and E.M.; methodology, E.M.; synthesis, E.M.; formal analysis, O.S. (Olga Sukhova); investigation, O.S. (Olga Sukhova), I.I., and O.S. (Olga Stonkus); data curation, M.K.; writing—original draft preparation, E.M.; writing—review and editing, E.M. and M.K.; supervision, Z.I. All authors have read and agreed to the published version of the manuscript.

Funding: This work was supported by the Ministry of Science and Higher Education of the Russian Federation within the governmental order for Boreskov Institute of Catalysis (project AAAA-A21-121011490008-3).

Institutional Review Board Statement: Not applicable.

Informed Consent Statement: Not applicable.

Data Availability Statement: Data is contained within the article.

Acknowledgments: The authors are grateful to G.S. Litvak, T.Ya. Efimenko, E.Y. Gerasimiv and V.A. Ushakov for their assistance with catalyst characterization. The TEM studies were carried out using facilities of the shared research centers: "National center of investigation of catalysts" at Boreskov Institute of Catalysis and "VTAN" at Novosibirsk State University.

Conflicts of Interest: The authors declare no conflict of interest.

References

1. Lupi, V.; Marsiglio, S. Population growth and climate change: A dynamic integrated climate-economy-demography model. *Ecol. Econ.* **2021**, *184*, 107011. [CrossRef]
2. Klein, T.; Anderegg, W.R.L. Global warming and urban population growth in already warm regions drive a vast increase in heat exposure in the 21st century. *Sustain. Cities Soc.* **2021**, *73*, 103098. [CrossRef]
3. Goeppert, A.; Czaun, M.; Jones, J.P.; Surya Prakash, G.K.; Olah, G.A. Recycling of carbon dioxide to methanol and derived products-closing the loop. *Chem. Soc. Rev.* **2014**, *43*, 7995–8048. [CrossRef] [PubMed]
4. Ismagilov, Z.R.; Parmon, V.N. Catalytic methods of processing carbon dioxide from coal generation into useful products. In *10 Breakthrough Ideas in the Field of Energy for the Next 10 Years*; Global Energy: Moscow, Russia, 2021; pp. 54–74. Available online: https://globalenergyprize.org/ru/10ideas/ (accessed on 25 June 2021).
5. Akaev, A.A.; Davydova, O.I. A mathematical description of selected energy transition scenarios in the 21st century, intended to realize the main goals of the paris climate agreement. *Energies* **2021**, *14*, 2558. [CrossRef]
6. Bulushev, D.A. Progress in catalytic hydrogen production from formic acid over supported metal complexes. *Energies* **2021**, *14*, 1334. [CrossRef]
7. Papadis, E.; Tsatsaronis, G. Challenges in the decarbonization of the energy sector. *Energy* **2020**, *205*, 118025. [CrossRef]
8. Bloomberg New Energy Finance. *Hydrogen Economy Outlook*; Bloomberg Finance L.P.: New York, NY, USA, 2020.
9. Cader, J.; Koneczna, R.; Olczak, P. The Impact of Economic, Energy, and Environmental Factors on the Development of the Hydrogen Economy. *Energies* **2021**, *14*, 4811. [CrossRef]
10. Dincer, I.; Acar, C. Innovation in hydrogen production. *Int. J. Hydrogen Energy* **2017**, *42*, 14843–14864. [CrossRef]
11. Mosińska, M.; Szynkowska-Jóźwik, M.I.; Mierczyński, P. Catalysts for hydrogen generation via oxy–steam reforming of methanol process. *Materials* **2020**, *13*, 5601. [CrossRef]
12. Chen, L.; Qi, Z.; Zhang, S.; Su, J.; Somorjai, G.A. Catalytic hydrogen production from methane: A review on recent progress and prospect. *Catalysts* **2020**, *10*, 858. [CrossRef]
13. Dawood, F.; Anda, M.; Shafiullah, G.M. Hydrogen production for energy: An overview. *Int. J. Hydrogen Energy* **2019**, *45*, 3847–3869. [CrossRef]
14. Le, V.T.; Dragoi, E.N.; Almomani, F.; Vasseghian, Y. Artificial neural networks for predicting hydrogen production in catalytic dry reforming: A systematic review. *Energies* **2021**, *14*, 2894. [CrossRef]

15. Mazhar, A.; Khoja, A.H.; Azad, A.K.; Mushtaq, F.; Naqvi, S.R.; Shakir, S.; Hassan, M.; Liaquat, R.; Anwar, M. Performance Analysis of TiO$_2$-Modified Co/MgAl$_2$O$_4$ Catalyst for Dry Reforming of Methane in a Fixed Bed Reactor for Syngas (H$_2$, CO) Production. *Energies* **2021**, *14*, 3347. [CrossRef]
16. Quarton, C.J.; Samsatli, S. The value of hydrogen and carbon capture, storage and utilisation in decarbonising energy: Insights from integrated value chain optimisation. *Appl. Energy* **2020**, *257*, 113936. [CrossRef]
17. Yu, M.; Wang, K.; Vredenburg, H. Insights into low-carbon hydrogen production methods: Green, blue and aqua hydrogen. *Int. J. Hydrogen Energy* **2021**, *46*, 21261–21273. [CrossRef]
18. Minutillo, M.; Perna, A.; Sorce, A. Green hydrogen production plants via biogas steam and autothermal reforming processes: Energy and exergy analyses. *Appl. Energy* **2020**, *277*, 115452. [CrossRef]
19. Worawimut, C.; Vivanpatarakij, S.; Watanapa, A.; Wiyaratn, W.; Assabumrungrat, S. Performance evaluation of biogas upgrading systems from swine farm to biomethane production for renewable hydrogen source. *Int. J. Hydrogen Energy* **2019**, *44*, 23135–23148. [CrossRef]
20. Chouhan, K.; Sinha, S.; Kumar, S. Simulation of steam reforming of biogas in an industrial reformer for hydrogen production. *Int. J. Hydrogen Energy* **2021**, *46*, 26809–26824. [CrossRef]
21. Khila, Z.; Hajjaji, N.; Pons, M.N.; Renaudin, V.; Houas, A. A comparative study on energetic and exergetic assessment of hydrogen production from bioethanol via steam reforming, partial oxidation and auto-thermal reforming processes. *Fuel Process. Technol.* **2013**, *112*, 19–27. [CrossRef]
22. Iulianelli, A.; Palma, V.; Bagnato, G.; Ruocco, C.; Huang, Y.; Veziroğlu, N.T.; Basile, A. From bioethanol exploitation to high grade hydrogen generation: Steam reforming promoted by a Co-Pt catalyst in a Pd-based membrane reactor. *Renew. Energy* **2018**, *119*, 834–843. [CrossRef]
23. Angili, T.S.; Grzesik, K.; Rödl, A.; Kaltschmitt, M. Life cycle assessment of bioethanol production: A review of feedstock, technology and methodology. *Energies* **2021**, *14*, 2939. [CrossRef]
24. Fu, J.; Du, J.; Lin, G.; Jiang, D. Analysis of Yield Potential and Regional Distribution for Bioethanol in China. *Energies* **2021**, *14*, 4554. [CrossRef]
25. Annual World Fuel Ethanol Production. Available online: https://ethanolrfa.org/statistics/annual-ethanol-production/ (accessed on 25 June 2021).
26. Nahar, G.; Dupont, V. Hydrogen production from simple alkanes and oxygenated hydrocarbons over ceria-zirconia supported catalysts: Review. *Renew. Sustain. Energy Rev.* **2014**, *32*, 777–796. [CrossRef]
27. Nahar, G.; Dupont, V. Recent Advances in Hydrogen Production Via Autothermal Reforming Process (ATR): A Review of Patents and Research Articles. *Recent Pat. Chem. Eng.* **2013**, *6*, 8–42. [CrossRef]
28. Sharma, Y.C.; Kumar, A.; Prasad, R.; Upadhyay, S.N. Ethanol steam reforming for hydrogen production: Latest and effective catalyst modification strategies to minimize carbonaceous deactivation. *Renew. Sustain. Energy Rev.* **2017**, *74*, 89–103. [CrossRef]
29. Sun, J.; Wang, Y. Recent Advances in Catalytic Conversion of Ethanol to Chemicals. *ACS Catal.* **2014**, *4*, 1078–1090. [CrossRef]
30. Hou, T.; Zhang, S.; Chen, Y.; Wang, D.; Cai, W. Hydrogen production from ethanol reforming: Catalysts and reaction mechanism. *Renew. Sustain. Energy Rev.* **2015**, *44*, 132–148. [CrossRef]
31. Chagas, C.A.; Manfro, R.L.; Toniolo, F.S. Production of Hydrogen by Steam Reforming of Ethanol over Pd-Promoted Ni/SiO$_2$ Catalyst. *Catal. Lett.* **2020**, *150*, 3424–3436. [CrossRef]
32. Greluk, M.; Rotko, M.; Turczyniak-Surdacka, S. Enhanced catalytic performance of La$_2$O$_3$ promoted Co/CeO$_2$ and Ni/CeO$_2$ catalysts for effective hydrogen production by ethanol steam reforming. *Renew. Energy* **2020**, *155*, 378–395. [CrossRef]
33. Olivares, A.C.V.; Gomez, M.F.; Barroso, M.N.; Abello, M.C. Ni-supported catalysts for ethanol steam reforming: Effect of the solvent and metallic precursor in catalyst preparation. *Int. J. Ind. Chem.* **2018**, *9*, 61–73. [CrossRef]
34. Vacharapong, P.; Arayawate, S.; Katanyutanon, S.; Toochinda, P.; Lawtrakul, L.; Charojrochkul, S. Enhancement of ni catalyst using CeO$_2$–Al$_2$O$_3$ support prepared with magnetic inducement for ESR. *Catalysts* **2020**, *10*, 1357. [CrossRef]
35. Sohrabi, S.; Irankhah, A. Synthesis, characterization, and catalytic activity of Ni/CeMnO$_2$ catalysts promoted by copper, cobalt, potassium and iron for ethanol steam reforming. *Int. J. Hydrogen Energy* **2021**, *46*, 12846–12856. [CrossRef]
36. Han, X.; Yu, Y.; He, H.; Shan, W. Hydrogen production from oxidative steam reforming of ethanol over rhodium catalysts supported on Ce-La solid solution. *Int. J. Hydrogen Energy* **2013**, *38*, 10293–10304. [CrossRef]
37. Moraes, T.S.; Neto, R.C.R.; Ribeiro, M.C.; Mattos, L.V.; Kourtelesis, M.; Ladas, S.; Verykios, X.; Noronha, F.B. The study of the performance of PtNi/CeO$_2$-nanocube catalysts for low temperature steam reforming of ethanol. *Catal. Today* **2015**, *242*, 35–49. [CrossRef]
38. Liu, Z.; Duchoň, T.; Wang, H.; Peterson, E.W.; Zhou, Y.; Luo, S.; Zhou, J.; Matolín, V.; Stacchiola, D.J.; Rodriguez, J.A.; et al. Mechanistic Insights of Ethanol Steam Reforming over Ni–CeO$_x$ (111): The Importance of Hydroxyl Groups for Suppressing Coke Formation. *J. Phys. Chem. C* **2015**, *119*, 18248–18256. [CrossRef]
39. Han, X.; Yu, Y.; He, H.; Zhao, J.; Wang, Y. Oxidative steam reforming of ethanol over Rh catalyst supported on Ce$_{1-x}$La$_x$O$_y$ (x = 0.3) solid solution prepared by urea co-precipitation method. *J. Power Sources* **2013**, *238*, 57–64. [CrossRef]
40. Cai, W.; Wang, F.; Zhan, E.; Van Veen, A.C.; Mirodatos, C.; Shen, W. Hydrogen production from ethanol over Ir/CeO$_2$ catalysts: A comparative study of steam reforming, partial oxidation and oxidative steam reforming. *J. Catal.* **2008**, *257*, 96–107. [CrossRef]

41. Matus, E.V.; Okhlopkova, L.B.; Sukhova, O.B.; Ismagilov, I.Z.; Kerzhentsev, M.A.; Ismagilov, Z.R. Effects of preparation mode and doping on the genesis and properties of Ni/Ce$_{1-x}$M$_x$O$_y$ nanocrystallites (M = Gd, La, Mg) for catalytic applications. *J. Nanopart. Res.* **2019**, *21*, 11. [CrossRef]
42. Ismagilov, Z.R.; Matus, E.V.; Ismagilov, I.Z.; Sukhova, O.B.; Yashnik, S.A.; Ushakov, V.A.; Kerzhentsev, M.A. Hydrogen production through hydrocarbon fuel reforming processes over Ni based catalysts. *Catal. Today* **2019**, *323*, 166–182. [CrossRef]
43. Kerzhentsev, M.A.; Matus, E.V.; Ismagilov, I.Z.; Sukhova, O.B.; Bharali, P.; Ismagilov, Z.R. Control of Ni/Ce$_{1-x}$M$_x$O$_y$ catalyst properties via the selection of dopant M = Gd, La, Mg. Part 1. Physicochemical characteristics. *Eurasian Chem. J.* **2018**, *20*, 283–291. [CrossRef]
44. Matus, E.V.; Ismagilov, I.Z.; Ushakov, V.A.; Nikitin, A.P.; Stonkus, O.A.; Gerasimov, E.Y.; Kerzhentsev, M.A.; Bharali, P.; Ismagilov, Z.R. Genesis and structural properties of (Ce$_{1-x}$M$_x$)$_{0.8}$Ni$_{0.2}$O$_y$ (M = La, Mg) oxides. *J. Struct. Chem.* **2020**, *61*, 1080–1089. [CrossRef]
45. De, S.; Zhang, J.; Luque, R.; Yan, N. Ni-based bimetallic heterogeneous catalysts for energy and environmental applications. *Energy Environ. Sci.* **2016**, *9*, 3314–3347. [CrossRef]
46. Dal Santo, V.; Gallo, A.; Naldoni, A.; Guidotti, M.; Psaro, R. Bimetallic heterogeneous catalysts for hydrogen production. *Catal. Today* **2012**, *197*, 190–205. [CrossRef]
47. Matus, E.V.; Ismagilov, I.Z.; Yashnik, S.A.; Ushakov, V.A.; Prosvirin, I.P.; Kerzhentsev, M.A.; Ismagilov, Z.R. Hydrogen production through autothermal reforming of CH$_4$: Efficiency and action mode of noble (M = Pt, Pd) and non-noble (M = Re, Mo, Sn) metal additives in the composition of Ni-M/Ce$_{0.5}$Zr$_{0.5}$O$_2$/Al$_2$O$_3$ catalysts. *Int. J. Hydrogen Energy* **2020**, *45*, 33352–33369. [CrossRef]
48. Kerzhentsev, M.A.; Matus, E.V.; Rundau, I.A.; Kuznetsov, V.V.; Ismagilov, I.Z.; Ushakov, V.A.; Ismagilov, Z.R. Development of a Ni–Pd/CeZrO$_2$/Al$_2$O$_3$ catalyst for the effective conversion of methane into hydrogen-containing gas. *Kinet. Catal.* **2017**, *58*, 601–622. [CrossRef]
49. Ismagilov, I.Z.; Matus, E.V.; Kuznetsov, V.V.; Mota, N.; Navarro, R.M.; Yashnik, S.A.; Prosvirin, I.P.; Kerzhentsev, M.A.; Ismagilov, Z.R.; Fierro, J.L.G. Hydrogen production by autothermal reforming of methane: Effect of promoters (Pt, Pd, Re, Mo, Sn) on the performance of Ni/La$_2$O$_3$ catalysts. *Appl. Catal. A Gen.* **2014**, *481*, 104–115. [CrossRef]
50. Moretti, E.; Storaro, L.; Talon, A.; Chitsazan, S.; Garbarino, G.; Busca, G.; Finocchio, E. Ceria-zirconia based catalysts for ethanol steam reforming. *Fuel* **2015**, *153*, 166–175. [CrossRef]
51. Trane-Restrup, R.; Dahl, S.; Jensen, A.D. Steam reforming of ethanol: Effects of support and additives on Ni-based catalysts. *Int. J. Hydrogen Energy* **2013**, *38*, 15105–15118. [CrossRef]
52. Akdim, O.; Cai, W.; Fierro, V.; Provendier, H.; Veen, A.; Shen, W.; Mirodatos, C. Oxidative Steam Reforming of Ethanol over Ni–Cu/SiO$_2$, Rh/Al$_2$O$_3$ and Ir/CeO$_2$: Effect of Metal and Support on Reaction Mechanism. *Top. Catal.* **2008**, *51*, 22–38. [CrossRef]
53. Moraes, T.S.; Rabelo Neto, R.C.; Ribeiro, M.C.; Mattos, L.V.; Kourtelesis, M.; Ladas, S.; Verykios, X.; Noronha, F.B. Ethanol conversion at low temperature over CeO$_2$-Supported Ni-based catalysts. Effect of Pt addition to Ni catalyst. *Appl. Catal. B Environ.* **2016**, *181*, 754–768. [CrossRef]
54. Pereira, E.B.; Homs, N.; Martí, S.; Fierro, J.L.G.; Ramírez de la Piscina, P. Oxidative steam-reforming of ethanol over Co/SiO$_2$, Co-Rh/SiO$_2$ and Co-Ru/SiO$_2$ catalysts: Catalytic behavior and deactivation/regeneration processes. *J. Catal.* **2008**, *257*, 206–214. [CrossRef]
55. Chen, L.C.; Lin, S.D. The ethanol steam reforming over Cu-Ni/SiO$_2$ catalysts: Effect of Cu/Ni ratio. *Appl. Catal. B Environ.* **2011**, *106*, 639–649. [CrossRef]
56. Kerzhentsev, M.A.; Matus, E.V.; Ismagilov, I.Z.; Ushakov, V.A.; Stonkus, O.A.; Larina, T.V.; Kozlova, G.S.; Bharali, P.; Ismagilov, Z.R. Structural and morphological properties of Ce$_{1-x}$M$_x$O$_y$ (M = Gd, La, Mg) supports for the catalysts of autothermal ethanol conversion. *J. Struct. Chem.* **2017**, *58*, 133–141. [CrossRef]
57. Kerzhentsev, M.A.; Matus, E.V.; Ismagilov, I.Z.; Sukhova, O.B.; Bharali, P.; Ismagilov, Z.R. Control of Ni/Ce$_{1-x}$M$_x$O$_y$ Catalyst Properties Via the Selection of Dopant M = Gd, La, Mg. Part 2. Catalytic Activity. *Eurasian Chem. J.* **2018**, *20*, 293–300. [CrossRef]
58. Li, D.; Nakagawa, Y.; Tomishige, K. Methane reforming to synthesis gas over Ni catalysts modified with noble metals. *Appl. Catal. A Gen.* **2011**, *408*, 1–24. [CrossRef]
59. Ji, H.; Cho, S. Steam-to-carbon ratio control strategy for start-up and operation of a fuel processor. *Int. J. Hydrogen Energy* **2017**, *42*, 9696–9706. [CrossRef]
60. Lee, S.H.D.; Applegate, D.V.; Ahmed, S.; Calderone, S.G.; Harvey, T.L. Hydrogen from natural gas: Part I—Autothermal reforming in an integrated fuel processor. *Int. J. Hydrogen Energy* **2005**, *30*, 829–842. [CrossRef]
61. Mikuli, E.; Migdal-Mikuli, A.; Chyzy, R.; Grad, B.; Dziembaj, R. Melting and thermal decomposition of [Ni(H$_2$O)$_6$](NO$_3$)$_2$. *Thermochim. Acta* **2001**, *370*, 65–71. [CrossRef]
62. Chen, K.; Zhang, T.; Chen, X.; He, Y.; Lang, X. Model construction of micro-pores in shale: A case study of Silurian Longmaxi Formation shale in Dianqianbei area, SW China. *Pet. Explor. Dev.* **2018**, *45*, 412–421. [CrossRef]
63. Zhao, P.; Qin, F.; Huang, Z.; Sun, C.; Shen, W.; Xu, H. Morphology-dependent oxygen vacancies and synergistic effects of Ni/CeO$_2$ catalysts for N$_2$O decomposition. *Catal. Sci. Technol.* **2018**, *8*, 276–288. [CrossRef]
64. Montoya, J.A.; Romero-Pascual, E.; Gimon, C.; Del Angel, P.; Monzon, A. Methane reforming with CO$_2$ over Ni/ZrO$_2$–CeO$_2$ catalysts prepared by sol–gel. *Catal. Today* **2000**, *63*, 71–85. [CrossRef]
65. Pan, Z.; Ding, Y.; Jiang, D.; Li, X.; Jiao, G.; Luo, H. Study on Ni-Re-K/Al$_2$O$_3$ catalysts for synthesis of N,N'-di-sec-butyl p-phenylene diamine from p-nitroaniline and 2-butanone. *Appl. Catal. A Gen.* **2007**, *330*, 43–48. [CrossRef]

66. Bobadilla, L.F.; Romero-Sarria, F.; Centeno, M.A.; Odriozola, J.A. Promoting effect of Sn on supported Ni catalyst during steam reforming of glycerol. *Int. J. Hydrogen Energy* **2016**, *41*, 9234–9244. [CrossRef]
67. Sharifi, M.; Haghighi, M.; Rahmani, F.; Karimipour, S. Syngas production via dry reforming of CH_4 over Co- and Cu-promoted Ni/Al_2O_3-ZrO_2 nanocatalysts synthesized via sequential impregnation and sol-gel methods. *J. Nat. Gas. Sci. Eng.* **2014**, *21*, 993–1004. [CrossRef]
68. Mondal, T.; Pant, K.K.; Dalai, A.K. Catalytic oxidative steam reforming of bio-ethanol for hydrogen production over Rh promoted Ni/CeO_2-ZrO_2 catalyst. *Int. J. Hydrogen Energy* **2015**, *40*, 2529–2544. [CrossRef]
69. Han, S.J.; Bang, Y.; Seo, J.G.; Yoo, J.; Song, I.K. Hydrogen production by steam reforming of ethanol over mesoporous Ni-Al_2O_3-ZrO_2 xerogel catalysts: Effect of Zr/Al molar ratio. *Int. J. Hydrogen Energy* **2013**, *38*, 1376–1383. [CrossRef]
70. Espitia-Sibaja, M.; Muñoz, M.; Moreno, S.; Molina, R. Effects of the cobalt content of catalysts prepared from hydrotalcites synthesized by ultrasound-assisted coprecipitation on hydrogen production by oxidative steam reforming of ethanol (OSRE). *Fuel* **2017**, *194*, 7–16. [CrossRef]
71. Muñoz, M.; Moreno, S.; Molina, R. Synthesis of Ce and Pr-promoted Ni and Co catalysts from hydrotalcite type precursors by reconstruction method. *Int. J. Hydrogen Energy* **2012**, *37*, 18827–18842. [CrossRef]
72. Profeti, L.P.R.; Ticianelli, E.A.; Assaf, E.M. Production of hydrogen via steam reforming of biofuels on Ni/CeO_2–Al_2O_3 catalysts promoted by noble metals. *Int. J. Hydrogen Energy* **2009**, *34*, 5049–5060. [CrossRef]
73. Gutierrez, A.; Karinen, R.; Airaksinen, S.; Kaila, R.; Krause, A.O.I. Autothermal reforming of ethanol on noble metal catalysts. *Int. J. Hydrogen Energy* **2011**, *36*, 8967–8977. [CrossRef]
74. Huang, L.; Zhang, F.; Wang, N.; Chen, R.; Hsu, A.T. Nickel-based perovskite catalysts with iron-doping via self-combustion for hydrogen production in auto-thermal reforming of Ethanol. *Int. J. Hydrogen Energy* **2012**, *37*, 1272–1279. [CrossRef]
75. Palma, V.; Ruocco, C.; Meloni, E.; Ricca, A. Oxidative steam reforming of ethanol on mesoporous silica supported Pt–Ni/CeO_2 catalysts. *Int. J. Hydrogen Energy* **2017**, *42*, 1598–1608. [CrossRef]
76. Cai, W.; Wang, F.; Daniel, C.; Van Veen, A.C.; Schuurman, Y.; Descorme, C.; Provendier, H.; Shen, W.; Mirodatos, C. Oxidative steam reforming of ethanol over Ir/CeO_2 catalysts: A structure sensitivity analysis. *J. Catal.* **2012**, *286*, 137–152. [CrossRef]

MDPI
St. Alban-Anlage 66
4052 Basel
Switzerland
Tel. +41 61 683 77 34
Fax +41 61 302 89 18
www.mdpi.com

Energies Editorial Office
E-mail: energies@mdpi.com
www.mdpi.com/journal/energies